Bird from HELL and Other Mega Fauna

Bird from HELL and Other Mega Fauna

Fourth Edition

Gerald McIsaac

Copyright © 2025 by Gerald McIsaac.

All rights reserved. No part of this publication may be reproduced, distributed, or transmitted in any form or by any means, including photocopying, recording, or other electronic or mechanical methods, without the prior written permission of the author, except in the case of brief quotations embodied in critical reviews and certain other noncommercial uses permitted by copyright law.

Printed in the United States of America

ISBN 979-8-9933020-3-4 (sc)
ISBN 979-8-9933020-4-1 (hc)
ISBN 979-8-9933020-2-7 (e)

Library of Congress Control Number: 2025921391

History

CONTENTS

Introduction vii
Part 1   Life In the Mountains 1
Part 2   Pterosaurs, AKA Pterodactyls 17
Part 3   Separate Species of Human 45
Part 4   Swimming Monsters 57
Part 5   Scientists and Scientific Theories 77
Part 6   Becoming Active 95
Part 7   Spiritual Power 103
Part 8   Raising the Level of Awareness of the Working Class 121
Part 9   Scientists: Form An Association 137
Part 10   Monopoly Capitalism 143
Part 11   The Industrial Revolution 155

INTRODUCTION

Recent developments have made it necessary for me to write a fourth edition to this book. For many years, I have been challenging numerous scientific theories, in various fields of science, including paleontology and anthropology. Hence the title of the book.

Yet it is also true that I am challenging theories in other fields, including economics and political science. For that reason, the title of the book is not completely accurate. But then, all fields of science are related. For that matter, no title is completely accurate, so I have chosen to stick with the above title.

As concerns my quest for species, which the scientists maintain to be extinct, my harshest critics point out that I have no proof that they exist. As if absence of proof is proof of absence! It is not!

Granted, all scientific textbooks, from those used in grade school to university, are agreed that around sixty-five million years ago, MYA, there was a mass extinction of dinosaurs. That is about the only thing upon which they agree! Different textbooks provide different definitions of dinosaurs! One textbook will maintain that dinosaurs were "land dwelling reptiles". Another will maintain that they were "active, intelligent, warm-blooded animals", while classifying flying reptiles, as "dinosaurs" Other textbooks provide other definitions of dinosaurs! They also disagree on the cause of that " extinction".

The most common theory is that of a "killer asteroid from deep space"! But now the theory of "extinction by continental drift" is gaining in popularity! No wonder children, as well as members of the public, are confused!

In addition, there is theory of the "mass extinction of mega fauna", here in North America.

Nonsense! Scientific fairy tales! Those mass extinctions never happened! There was no mass extinction of dinosaurs, just as there was no mass extinction of mega fauna!

Yet my critics maintain that those extinctions must have happened, so many species *must* be extinct, because *all scientific textbooks agree that they are extinct!*

I maintain that a great many species, which the scientists maintain are extinct, are very much alive. I say this, with the utmost confidence, as the people with whom I have lived, for over forty years, have described these animals to me. This calls for a little explanation.

Perhaps a little background is in order.

Allow me to start by saying that I am a Caucasian male with a scientific background. Over forty years ago I got tired of big city life and opted for life in the mountains. It is not a decision I regret. On the contrary, it is a lifestyle I love.

I have since married an Indigenous girl, a member of the Sekani tribe, of the Dene people. We have raised a family on a Reserve referred to as Tsay Keh Dene, or Mountain People, loosely translated. The village is located in British Columbia, in the Rocky Mountain Trench, 350 kilometres or 200 miles from the nearest town, that of MacKenzie. The closest neighbouring Reserve is that of Kwadacha, 75 kilometres or 50 miles away.

Allow me to stress the fact that the Mountain People are the experts on the mountains! The one and only way they could have possibly survived in this unforgiving wilderness, is by having an intimate knowledge of the mountains, and the animals within these mountains. For that reason, I believe the Mountain People. I do not believe the scientists! As the scientists and the Mountain People are saying precisely the opposite things, it is simply not possible to believe both!

Out of respect for our American neighbours, most of whom are not familiar with metric measurements or of Celsius, I have chosen to provide distances in imperial as well as metric, and temperatures in Fahrenheit as well as Celsius. Further, that which is referred to as a Reserve in Canada, is referred to as a Reservation in America.

As most people who live in the United States refer to themselves as Americans, I have chosen to use that terminology, at the same time referring to the country as America.

The social structure on the Reserve is different from that to which most people are accustomed. Without going into any great detail, suffice it to say that such a social structure gives rise to very strong social bonds. Also, the names of the deceased are not to be mentioned, so I do not refer to them directly.

The older members of the Band, as the community refers to itself, are referred to as "Elders", and are regarded with the utmost respect. They are the equivalent to those referred to as "Seniors" in conventional society.

The Elders are the absolute experts on the mountains, and I use the word "mountains", in the loosest possible sense. This is to say that I include the forests, meadows, swamps, lakes and rivers. The Elders have provided me with detailed descriptions of animal which I recognize. The same animals which the scientists maintain are extinct!

I maintain that the Indigenous Elders are correct, and the scientific community is completely mistaken. Now it is the not so little matter of proving this.

There is a sense of urgency in this, and not only because we are entitled to our wildlife. It is part of our heritage. Also, because many of these animals are predators. All carnivores have a keen sense of smell, as do most predators, and many of these animals are predators. Further, all predators are far more likely to attack when they smell blood. That is the reason they prey mainly upon girls of childbearing age. It is absolutely essential that all members of the public be made aware of their existence.

It is correct to regard this as a book of science, which it is, one which is written in a popular manner, complete with some rather poor jokes. This stands in contrast to most books of science, which few common people read, if only because they cannot understand them. Yet as the scientists are not performing their duty, including that of documenting the existence of these animals, then it is up to common people to take up the slack. Of necessity, certain technical terms are mentioned and properly explained.

I can only hope that a great many people will be motivated to assist in locating these magnificent animals. You will not be disappointed.

PART 1

LIFE IN THE MOUNTAINS

Mountain People

The Dene Elders grew up in an environment which was very similar to that which is commonly referred to as the Stone Age. In fact, it was a time of great change, of cultural conflict, as it is now. They were being introduced to civilization, complete with modern tools and weapons, as well as diseases. They were also expected to adapt, to change their social and cultural way of life, in order to fit into civilized society. That same civilized society had no intention of adapting to the lifestyle of Indigenous People. In fact, neither culture made any great effort to understand the other.

The Dene Elder were members of a hunting gathering society. It is an understatement to say that life was difficult. All food had to first be gathered or killed. All tools had to first be fashioned. They lived constantly on the edge of starvation.

That changed dramatically, at the time when the Dene first met traders from civilization. Those traders offered burlap sacks of flour, rice and sugar, in exchange for peltries, otherwise known as furs. This is to say that the traders were interested in the hides of animals, such as marten, mink, beaver, lynx, wolf, wolverine, coyote, grizzly and such.

The Dene had their own priorities. They knew how to weave baskets, and used these baskets, especially for gathering berries. These baskets were quite useful, but rather feeble. They were not capable of carrying anything much heavier than berries. By contrast, the rice, flour and sugar came in containers of burlap bags. As burlap is very strong, the Dene traded peltries for burlap bags. They had no interest in the contents, if only because they did not recognize it as food. The flour and rice were thrown on the ground, while the sugar was thrown into the fire. They liked to see it sparkle as it burned. Entertainment!

Further, the Dene were at such a primitive stage of development, that they had no pottery. For that reason, the mere act of boiling meat was difficult. The only way in which this could be done, was by using the stomach of an animal, as a container. The water was added to the container, hot rocks from the campfire were tossed into the water, and the meat was added. Mountain goat was considered the best for this, as they were able to get three meals, from each container, in this manner. This should give some idea of the world in which the Dene lived.

The resultant cultural confusion is understandable and persists to this day. Of course they were nomads, following the herds, unable to stay in one place for any length of time, if only because their food supply would soon be exhausted. Their survival depended largely upon their knowledge of the animals, and their habits. The fact that they survived, is proof that they are the experts, in regard to these mountains. For that reason, I pay strict attention to that which they tell me.

Numerous people have questioned the sanity of the Dene. They point out that those who choose to live in such a harsh environment, must be crazy. To such people, I can only respond that there was not a great deal of choice in the matter.

It is the scientific opinion that there were possibly three migrations of people across the Bering Strait, from Asia, over a period of possibly thousands of years. The Rocky Mountain Trench provides a natural corridor. The ancestors of the people with whom I live, apparently came across the Strait in the most recent migration, possibly around the time of Columbus, within the last few hundred years.

These "immigrants" soon found themselves in a situation which is similar to that which modern day immigrants face. In other words, "not welcome"!

The Dene immediately found that the finest "real estate", such as the coastal regions, was already occupied. Such regions provided the inhabitants with a great abundance of seafood, including fish, whale, walrus, seal, sea lion and dolphin. Also, freshwater fish, as well as deer, moose, elk, bear, duck, goose and so forth. In addition, the surrounding forests of cedar and hemlock provided the material for lodging and tools. Such coastal areas were as highly prized then, as they are now. Further, the people who lived in those areas were not at all anxious to "share their wealth".

That left the land to the east, the prairies, with vast herds of bison and other wildlife, but that too was occupied. Those people too, had no intention of sharing. The Dene, as the latest immigrants, had no choice but to subsist, as best they could, in the corridor, the Rocky Mountain Trench.

The Trench is famous for its magnificent scenery and big game. Professional photographers have taken countless pictures of the mountains, while trophy hunters are enthusiastic about killing the biggest male animals, such as moose, elk, caribou, mountain sheep, mountain goats and bears. Grizzly and mountain sheep are the most prized trophy animals, and the hunters pay a small fortune, for a chance to kill one of those animals.

The Dene who lives in the mountains are more impressed by the difficulty in surviving. The mountains are truly magnificent, as the poets and photographers can testify. It is also a fact that those same mountains are not the slightest bit forgiving. Any moment of complacency can prove to be fatal.

Recently, a couple of girls were reminded of this. They decided to head for town in the middle of winter, with the temperature close to forty degrees below zero. At that temperature, Fahrenheit and Celsius are almost identical, supremely cold. Their vehicle broke down and they had no two-way radio, so no way to call for help. That and the fact that they had no matches, no way to light a fire, meant that they were in serious trouble. They spent a severely cold night in the pickup and were found the next day. They could not have survived another night in that severe cold.

The only reason they were not rescued immediately, was because no one was looking for them. The first vehicle that came along stopped and offered assistance. In the mountains, this is common practice. We tend to take care of each other. After all, the roads in these mountains are not to be confused with highways.

Those two girls lived in these mountains all their lives, and had driven those roads many times, without any major mishap. As a result of this, they developed a certain complacency, which nearly cost them their lives. Of course, they should have carried with them a box of matches and candles, as well as a two-way radio. Under such circumstances, the vehicle provides the shelter, and the lit candles provide sufficient heat, to keep body and soul united.

It was only within the last thirty years, that roads and bridges have been built, connecting the village with the highway. Electricity is supplied by generators, and the village now has running water, a modern school, clinic and store. This has resulted in a vast improvement in the living standards of the people who live there.

The roads leading into the village are maintained by the Forest Service. These are commonly referred to as "dirt roads", while in fact they are gravel roads. They are also sometimes referred to as logging roads, although gravel trucks, mining ore trucks and low bed trucks also use them. Such vehicles are referred to as "trucks", while passenger vehicles are referred to as "pickups". Very few cars travel these roads, as they cannot take the punishment.

These roads tend to be rather narrow, with the occasional "wide spot" that are suitable for pulling into, when approached by a "wide load". A loaded logging truck or ore truck, as well as a low bed truck, carrying a big machine, qualifies as a "wide load".

At the start of each road, signs are posted. These signs state the name of the road, as well as the correct radio frequency to be used. As well, kilometre signs are posted every kilometre or two. Those who travel these roads are expected to have a two-way radio, to announce their location and direction of travel, on a regular basis, perhaps every couple of kilometres. Vehicles which are "loaded", which generally means going in the direction of town, have the right of way. It is up to vehicles which are "empty", or going in the opposite direction, to get out of their way.

Until quite recently, the various "Main Lines", or logging roads, were assigned certain radio frequencies, and these frequencies were assigned various names. As these names and frequencies varied across the province, it led to a certain amount of confusion. When driving these roads, the last thing anyone needs is confusion. That can cost someone their life.

For that reason, the Forest Service decided to change the system, making it standard. Now each radio frequency is given a number, and these numbers are used throughout the province. This simplifies the matter but also meant that everyone had to buy a new radio. It is a small price to pay for safety. As well, the words loaded, and empty have been replaced with "up" and "down".

As an example, a logging truck travelling south to the mill, with a load of logs, on the Finlay Main Line, may see a sign that reads "18km". All drivers know that km is short for kilometres, so the driver is expected to get on the two-way radio and announce "eighteen down on the Finlay". If the driver is in a pickup, then that should also be specified. Any vehicles going in the opposite direction, "up vehicles", are expected to pull over into a wide spot, to allow the "down vehicle"' to pass. It should be noted that people who use these roads, tend to refer to kilometres as "clicks".

As for those who consider this to be somewhat unfair, that a loaded truck has the right of way, consider the fact that each logging truck may carry fifty tons of logs. If the truck makes a sudden stop, then the logs it is carrying tend to come forward into the cab of the truck. Such things do happen, and they tend to ruin the whole day of the driver.

The side roads, which branch off the various main lines, lead to logging areas, which we refer to as "logging blocks". After the timber is harvested, the roads inside that block are "deactivated".

This means that a machine, usually a backhoe, goes into the area and tears up the road at regular intervals. As well, the bridges are usually removed. This generally happens only after the block has been replanted.

The road between the two Reserves of Tsay Keh Dene and Kwadacha, is called the Russel Main Line. It follows the Finlay River, but on the west side of the river. On the east side of the river, there is another Main Line, called the Finlay Main Line. Traffic on the Russel Main Line uses a different radio frequency than traffic on the Finlay Main Line. In each case, the vehicles are expected to specify which road they are travelling. This is in the interests of safety, as we all make mistakes. "Down" vehicles are especially expected to announce their location, while "up" vehicles may also announce, if only less frequently.

No doubt, there are a great many people who have a difficult time imagining life in this remote area. Those who first move up here, especially teachers, tell me that the culture shock is severe. It is similar to moving halfway around the world! The point being that if the reader has a difficult time in imagining life in these mountains, that is completely understandable. For that reason, I am going into such detail.

These mountains are rugged and remote and are home to several species of animals that are thought to be extinct. They are rarely seen, if only because there are so few people living in these mountains.

Many years ago, I became convinced of the existence of these animals and decided to devote my life to proving this. My method involves separating fact from belief. I focus on the description of the various animals, which tends to be detailed and accurate. At the same time, I set aside the personal beliefs of the individual, who is describing the animal. I respect the beliefs of all common people, members of the public, while not necessarily sharing those beliefs. I have my own beliefs. In turn, I expect all others to respect my beliefs.

MASS EXTINCTION OF MEGA FAUNA

As previously documented, it is the scientific opinion that possibly ten thousand years ago, at the end of the last ice age, there was a "mass extinction of mega fauna", in that five species of huge animal dropped dead, because they were unable to handle climate change. These five species include the woolly mammoth, the Jefferson ground sloth, the dire wolf, the sabre-toothed cat and the short faced bear. In fact, all of these animals are truly huge, or mega. What is more, they are not extinct.

For the benefit of those who are not terribly familiar with scientific terms, I will mention that "mega" means huge while "fauna" means animals. I will also mention that this theory, that of their mass extinction, is "mega" ridiculous!

As all scientists are supremely well aware, within the last hundred thousand years, we have had no less than three ice ages, here in North America, all of which the "mega fauna" survived. Those same scientists are also well aware that each time the climate changed, which is to say that each time the continent cooled off and glaciers covered the land, these species survived. That merely stands to reason. It also stands to reason that each time the continent warmed up and the glaciers melted, the species also survived. The fact that these species were alive at the end of the last ice age proves this, beyond any shadow of a doubt. It also proves, also beyond any shadow of a doubt, that these species are quite capable of handling climate change. Further, they did. Those species are still very much alive.

I became convinced of the existence of the woolly mammoth, when my mother-in-law recognized a picture of an elephant. The only difference is that the "elephant" she remembered was "hairy". That is the reason the Dene refer to this animal, the woolly mammoth, as the "hairy elephant". This is to say that *the woolly mammoth still exists!* The Dene were running from the woolly mammoth in the late twentieth century! Their one and only chance against this animal, one of the largest land-dwelling animals in the world, was to make it to the safety of the nearest swamp!

As the swamp represents muskeg, and the animal senses that it has to avoid muskeg, the people were safe in the swamp. In winter this is not an issue, as the mammoth spends the winter in caves.

This is not to say that the mammoth is a flesh eater, because it is not. It is to say that the mammoth is almost certainly an intelligent animal, and as such, is very likely capable of emotion. Assuming that to be true, then it is clear that the mammoth feels emotion. The emotion it feels towards us is that of hatred. Who can blame it? Until very recently, the mammoth was widespread across North America.

It was especially at home on the prairie. Then, at the time of the European invasion, settlers appeared. The last thing a settler wants on his homestead is a five-ton vegetarian. They made this quite clear by shooting every mammoth they could get in their sights.

The modern-day farmers of Africa are currently in a similar situation. Their only livelihood is their crops, and the elephants take great delight in feasting and trampling the crops which the farmers have worked so hard to grow. But then those crops are generally far superior, far more nourishing, than the vegetation which grows naturally in the area. No doubt, the farmers who are suffering the loss of their crops would love nothing more than to kill every elephant which comes close to their farm. They are unable to do so due to laws. The elephants are most valuable, if only for purposes of attracting tourists.

The American homesteader of the late nineteenth century was not restricted by such laws. As the railroads pushed west from the province of Ontario, they carried settlers who cultivated the land and planted their crops. They also protected those crops, with firearms, when necessary. There was no law against killing the woolly mammoth.

Those animals responded by running ever further west and north, into the relative safety of the mountains. The remnants of a once great herd, which at one time stretched across North America, is now scratching out a living in the unforgiving mountains.

The problem now is to prove they exist, pass laws to protect them, and lead them out of the mountains to their former grazing grounds. Wildlife photographers will line up to take their pictures.

The farmers on the prairie will not be so enthusiastic to see these mammoths. The mammoth will find themselves in a land transformed, a "land of milk and honey". They will marvel at fields of wheat, rye, corn and oats, as far as the eye can see. The woolly mammoth can see very far! They will then walk through the fences of the farmers and have a feast. The difference now is that the farmers will not be allowed to kill the woolly mammoth, as they will be protected by law, just as the elephants of Africa are protected.

The government will have to find some way to compensate the farmers for the loss of their crops, perhaps by charging a fee to tourists who delight in taking pictures of these magnificent animals. No doubt in the fall, instincts will kick in which force the mammoth to migrate south to warmer pastures, passing through the international border.

Once the animal gets the idea that it is safe, that we are not going to shoot it, it may in turn do its best to irritate us, just to "get even". It may block roads and highways, trample golf courses, even knock down fences, just to let us know that it hates us. We had best be prepared.

As the animal is so huge, locating it and proving it exists should not be difficult. It can possibly be seen from satellites, when there is no cloud cover. It can certainly be seen from small planes. The pilots who fly across these mountains, referred to as "bush pilots", no doubt see this animal on a regular basis.

Equally without doubt, certain forest service personnel, those who fly across the forest districts looking for forest fires, also see the mammoth. They are careful to report the location of any and all forest fires, but equally careful to not report the location of any mammoth. Their careers depend on this.

This brings us to another species of "mega fauna", also allegedly extinct. This animal is commonly referred to as the "short faced" bear, technically referred to as "arctodus simus". The Dene refer to it as the "rubber faced" bear or the "beaver eating" bear.

The bear is truly huge, weighing in at a ton. That means one thousand kilos or two thousand pounds. It stands one point eight meters or six feet at the shoulder. When it rears up on its hind legs it stands five meters or seventeen feet high. The claws are seven inches or fifteen to eighteen centimetres long. They use these claws to tear apart beaver houses!

As far as I am aware, no bear has ever been documented to tear apart a beaver house! For that matter, I have never heard of any bear that has even tried to tear apart a beaver house, if only because they are able to sense that they cannot do this. Yet this bear makes a habit of tearing apart such beaver houses. It is that big and powerful!

At least in spring, when the beaver pond is still frozen, the short faced bear tears open the beaver houses. Of course, the beavers dive into the water, but as they still have to breath, come to the surface in the only opening in the ice, right into the jaws of the short faced bear.

To clarify, the Dene tell me that the bear has no hair on its face, so that is the reason they call it the rubber faced bear. The reason they call it the beaver eating bear is quite obvious. I have included two stories of encounters with the short faced bear, by a Dene elder.

As I consider the stories so important, I have decided to include them here, by permission of that elder, Seymour Isaac. These incidents happened at the time when he and his brother Francis were lads. The men were teaching them how to hunt beaver, so that the boys very likely had twenty-two calibre rifles, while the men had high powered rifles, or "big guns", as they call them. I have included the story as it was written by Elder Seymour.

I should add, for the benefit of those who are not familiar with imperial units, that "six or seven inches" is the equivalent of fifteen or eighteen centimetres. Also, seventeen feet is the equivalent of approximately five metres.

But now let the elder tell his story, in his words: The title is Beaver Eating Bears:

"I remember the first beaver eating bear that was shot and killed way back in 1953 or 1954. Grandpa Keom Pierre and his stepson Larry Pierre and William Poole and Jimmy Dennis, Charlie and I, who were all young boys at the time, not yet in our teens, were trapping beaver. We had been doing so for five days and headed up to Shovel Creek Pass on the far side of the trap line which Grandpa Keom called 15 Mile Cabin. Around the first week of April, there was still four to five feet of snow, so we used snowshoes. We had set beaver traps under the ice and found some places that had exposed spring water.

"We knew that Davis Creek flowed past 15 Mile Cabin only about three miles away. Keom and Larry also knew that the Davis always opened up early, so the boys were told to bed early. At first light, we would head to Davis Creek while the snow crust was still frozen. The next morning, we all got ready to travel on top of the crusty snow and took three of Grandpa's hunting dogs with us. The dogs were named Tez, Pup and Silver. We travelled on and finally came to the creek. The dogs started barking a commotion like they had seen something they did not like -and they certainly did. In front of us stood the biggest bear we had ever seen; it was bigger than the biggest bull moose we had ever seen. Then the fun began.

"They started shooting and Keom just shouted, 'Don't hit the dogs,' but the dogs had the advantage, because the bigger bear was breaking through the four-inch snow crust, and the dogs were on top of the snow and not sinking through. Since the bear was breaking through the crust, it could not move fast. Once that bear saw us, the bears concentration was on the dogs. The bear tried to get them, and the men tried to move away from the bear as they were shooting. After twelve or thirteen well place shots, the bear finally dropped. Grandpa Keom had used an 8-millimetre war gun, Uncle Larry had used a 303 British war gun, and Jimmy fired his brand new 30-30 Marlin lever action. But it was William, with a single shot .22, who took the bear down that day.

"While the bear was being shot at, it thrashed everything and anything that was in its way. It ripped out little trees; willows flew left and right. More than a few times, the bear stood up and was a massive seventeen feet tall. Man, did he stink! After the chaos was over, the bear was skinned. Its claws were six to seven inches long. The hide could not be saved because the side that he slept on was ruined from his shoulder to his hips from rubbing."

The second story from this Elder is titled "Beaver Eating Bears in Akie":

"In 1958, my brother Francis Isaac and I were up the Akie River to do some spring beaver trapping with Uncle Angus Pierre and Uncle Mac Pierre and the whole family. We just loved Grandma Elizabeth and the great stories that she always told.

"At about mid-May, the five of us being myself, Uncle Angus, Auntie Lucy, Uncle Mac and my brother were going up the Akie. Old Grandma was left to babysit at the time.

"We started to travel up the river. As I recall, the Akie is a very difficult river to shoot beaver on in the spring. At that time of the year, the river is so low it almost becomes a creek. When the river is low, the beavers do not come out until dark, so we mostly used traps. We did pretty good in trapping on a daily basis and trapped three or four beaver each day. When we baited the traps, we used poplar tops mixed with beaver castor and used a little of our own oil mix, which worked very well in the traps as bait. Since we camped along the way, we would set our traps and enjoy ourselves until it was time to check the traps.

"As we cut trail along the way, Uncle Mac told us to look out for ourselves as there was a lot of bears and grizzlies in the area, especially where we were cutting trail. During our way along the trails, we encountered bears and scared them away. We didn't encounter any problems until that one fine day on the way back. We passed the second trapping cabin and went right on up to the area that's called the big bend in the Akie River.

"Our Sekani ancestors had a name for that place and called it Big Bag Yorks. Uncle Mac shared some history of that place. Our people used to stop there to dry meat, and the women picked berries and medicinal plants. Big Bag Yorks was a nice place to camp. It got its name by being a place where most of the valleys could be seen. Creeks ran into the Akie, and there was plenty of game in that area. This place should be marked as a historical marker on our map as a sacred place.

"We had camped in the area for about three days before heading down the Akie. On our third day of camping, we encountered what our ancestors called the beaver hunter.

"At the time, Uncle Mac was setting a beaver trap. Uncle Angus told him that we would travel on and wait down at the point for him. We came to a place with a lone sandbar right by the bluff. The water must have been quite deep as I remember an eddy in the green, pristine water.

"In that area, we saw a beaver jump in. Uncle Angus took out his gun and went down the bluff to look for that beaver. Right across the river, we saw a dry slough that ran right into the Akie. I thought we had seen a moose and Francis thought that too. All of a sudden, Francis looked at Uncle Angus and wanted to know who was on the sandbar? At first, we thought maybe it was someone, but it was a beaver hunting bear coming towards us on the sandbar. It was only fifty feet away from us. Uncle Angus was ready and grabbed his gun and shot. Those first two shots were so fast and accurate that it sounded like one shot! Uncle Angus shot four more shots then ran and knelt by a tree. He started shooting more. Although Uncle Angus had a single shot 30-30, he was shooting out shell like an automatic. Oh, the action! In the meantime, when all this shooting was going on, we could hear that beaver hunting bear yelling and hollering, crawling, rolling around, and scooping up paws full of gravel. The huge bear stood up. After fifteen shots, it had to have felt the shots and started walking on the little island. One 30-30 and one 303 British had hit him. Finally, we heard his last, big growl. Some of us wanted to check the bear out, but Uncle Angus noted it was getting late, and we knew it was dead.

"The next day, we got up early and headed back home. We told old Grandma and the kids about the encounter with the beaver eating bear. She told us that she knew about it. As long as everyone was okay, she was happy.

"Between the four of us, in twelve days we got forty-five beaver and twelve muskrats!

"After we were back, Billy and Art Van Somers came up and brought supplies. They brought a parcel and a letter for me from Dad saying he wanted both of us back to town. He also mentioned he had missed us very much and that he was staying with our sister Louise and her husband Wilson in Summit Lake. When they came back, we went to town with them and shared our proud hunting story."

To continue with our list of mega fauna, the Jefferson ground sloth is named after Thomas Jefferson, the third president of the United States. The Dene refer to this animal as the giant beaver.

In all fairness to Thomas Jefferson, the man was a genius, truly talented. As well as being a diplomat and a politician, he was a scholar, inventor, scientist and architect. He personally designed the building on his plantation of Monticello. It is considered to be an absolute masterpiece. This was the man who first described the animal which is named in his honour, the Jefferson ground sloth.

This in no way changes the fact that the man was a slave owner, a true psychopath. Nor does it change the fact that the scientists are supremely well aware that the animal, which was named in his honour, was alive two hundred years ago, as it is today. Yet to this day the scientists insist that the animal is extinct!

Another example of the hypocrisy of the scientists is the supposed extinction of the dire wolf. The wolf is huge, the size of a deer. It may weigh perhaps one hundred fifty pounds or seventy kilos. The Dene refer to this animal as the "wilderness wolf". It is well named, indeed "dire", as this animal is not afraid of us! They come into the village after sundown and prey upon dogs. Those dogs which are tied up are particularly easy prey, for these wolves.

Even the scientists admit that the dire wolf still exists, even though they claim that it is extinct. As they phrase it, the dire wolf exists, but only as "a remnant population of an extinct species". Such nonsense is now referred to as "alternative facts", although I have another name for such foolishness. I call it what it is. A pack of lies!

This brings us to the sabre toothed cat, technically referred to as Smilodon fatalis. This huge cat, the size of a Siberian tiger, loves the grass land, so is not located in the mountains. It loves the prairie, and in fact has been seen on numerous occasions within the city limits of Milwaukee. For that reason, it is commonly referred to as the "Milwaukee lion". There are numerous pictures and videos of the animal, posted on the internet. It is a female, which explains the reason the animal lacks huge teeth. Such teeth are characteristic of the males of the species only. The males have evolved those "sabre teeth" for purposes of display.

PART 2

PTEROSAURS, AKA PTERODACTYLS

The "Devil Bird"

If there is one thing that terrifies the Dene Elders, it is the animal they refer to as the "Devil Bird". It is their belief that it is an evil spirit, a demon, a "bird from hell".

This belief is based on the fact that the animal is nocturnal. It spends the day light hours inside caves. It comes out of those caves only after sundown, and hunts in open areas. It also tends to avoid any light.

Incidentally, I have since determined that this is the same animal which is referred to as the "Dragon", in many other parts of the world. The fact is that the Chinese have named the twelve years after twelve animals, and that includes the Dragon. That is referred to, quite reasonably, as the "Year of the Dragon". I am convinced that the Dragon exists, and very soon, will prove this.

It is also a fact that the Dragon is mentioned in the bible. As the bible was written in Africa, this means that the Dragon is in Africa, as well as Asia.

It is only in the New Testament, within Revelations, that the Dragon is identified as the "Beast", or "Satan". That is precisely the belief of the First Nations people!

It is not just the Dene Elders who believe that this animal is a spirit. It is safe to say, that all across North America, First Nation Elders can swear to the fact that these animals exist. Most of them just think that it is a spirit, and not an animal. We must respect that belief.

Countless people are aware of the belief, of the Elders, in "spirits", yet most of those people consider this to be nothing more than a "myth". They could not possibly be more mistaken! But then none of them has thought to ask the Elders to describe this "Devil Bird".

That did not stop me from asking that question, as I am not terribly shy. The answer they provided, shocked me, to the very core of my being: "It has a head like an eagle, the body is the size of a man's body, the wings are as wide as two moose hides stretched out, the flapping of the wings sounds like dry hide, and the tail is as long as a man is tall, and it ends in the shape of an arrowhead".

This is a near perfect, accurate, detailed description of a pterosaur, an animal the scientists have classified as a dinosaur, one which has been *extinct for sixty-five million years! Or not!*

A little explanation is in order. This animal is a flying reptile, and the correct scientific term is that of pterosaur, although they are also known as (AKA) pterodactyls. The majority of these flying reptiles have long tails, and the technical term for them is "rhamphorynchus". The flying reptiles with short tails are less common and are classified as the true "pterodactyls".

Most common people have heard of pterodactyls, but not of pterosaurs or rhamphorynchus. Bear in mind that they are commonly referred to as dragons, although local names include Thunder Bird and Jersey Devil, among others.

The wingspan of these huge flying reptiles is estimated to be between ten and fifteen metres, or thirty-five to fifty feet. This is three to four times wider than the wingspan of the largest flying bird in the world, the albatross. Further, all four limbs are involved in flapping the wings. This also gives it great lifting power. In fact, it is able to pick up deer and pack them away! Quite an accomplishment, as deer weigh around 90 kilos, or 200 pounds.

The Dene Elders are aware of this, as are no doubt the Elders all across the country. For a fact, the Dene Elders do not go outside after sundown. They know better! As they put it, "a big bird will pick you up and carry you away!"

In fact, in a spirit of cooperation, they have pointed out to me, certain caves in the mountains. These caves open up onto a vertical face, otherwise known as a cliff or a bluff, or "straight up". The tops of these mountains are horizontal, or "flat", to put it in popular terminology. For that reason, I refer to these mountains as "perpendicular mountains".

This is significant, as these mountains form the "nesting ground" of the pterosaurs. With their distinctive shape, they are easily spotted. The point being that it is easier to locate the nesting ground of the pterosaurs, than it is to locate the animal itself.

Incidentally, the Elders tell me that those caves lead directly to hell. They further tell me that after sundown, Satan opens the gates of hell and releases his demons, his "birds from hell"! They know this because they hear a snapping sound!

Allow me to stress the fact that I respect their beliefs, just as I respect the beliefs of all common people. Which is not to imply that I share all of those beliefs.

This stands in sharp contrast to the facts they present to me. One of those fact is that after the animals comes out of those caves, it does not just fly away. Instead, it climbs straight up that vertical face.

As for those who are skeptical -and I hope there are a great many of you! - allow me to say that I welcome such an attitude. I just wish the scientists had that attitude!

The explanation, for the fact that these flying reptiles, pterosaurs, are able to climb straight up a cliff, can be found in the fossilized track ways of the pterosaurs! I mention this as an example of combining the wisdom of the Elders, with scientific facts.

The point is that the scientists have discovered the fossilized trackways of the pterosaurs. These trackways tell us a few things about the animals, things which the fossilized bones and eggs, of the animals, cannot tell us. They tell us that the animal sometimes walks on two legs and sometimes walks on all fours! This is to say that the wings of the animal, double as legs! The wings are used for walking, as well as flying! The pterosaurs climb up that cliff on all fours!

I should add that, once it gets to the top of the mountain, at the point where it is flat, it no doubt opens its wings. That explains the snapping sound.

This also explains the existence of the "devil dog", as the animal is called, when it walks on all fours. Of course, it is called the "devil bird", when it walks on two legs.

This is an example of separating the facts, which the scientists present, from the theories they present, even though they present those theories as facts! The facts are just that. Facts! Their theories, such as the belief that the pterosaurs are extinct, are not based upon any scientific facts.

Absence of proof is not proof of absence! Just because we have no proof that they still exist, at least not yet, does not mean that they are extinct! To present a theory, based on no facts, is completely contrary to the scientific method!

Now to return to the subject of the pterosaurs.

It should be stressed that this animal is nocturnal, which is to say that it avoids the light, as much as possible. It comes out of those caves after sundown and returns to those same caves before sun rise. That is the reason it is so rarely seen.

The entrance of those caves may, or may not, be very high off the ground. It is necessary only for those caves to be high enough that predators, such as bears and wolves, cannot climb into the caves. This is to say that four to five metres, or fifteen to twenty feet above the ground, is sufficient.

Incidentally, numerous people have asked me if these animals can be spotted on radar. To this question, I can only respond that the question is backwards. As they are the size of small planes, the real question should be: *How can they not be seen on radar?* More on that subject, later on in the article.

Now to proceed with our investigation.

Pterosaurs Reproducing

These animals are reptiles, so that they reproduce by laying eggs. Further, reptiles cannot generate their own body heat. The fact is that the eggs of all reptiles have to receive heat from an outside source, as otherwise they cannot hatch. The eggs of crocodiles receive their heat from rotting vegetation, for example. Turtles lay their eggs in the sand, at the edge of the water, either lakes or the ocean. The sun warms up the sand, along with the eggs. But where do the eggs of pterosaurs receive their heat?

Certainly not from inside the caves, as the temperature inside those caves is roughly a constant 13 degrees Celsius, or 57 degrees Fahrenheit. That is too cold for even the eggs of reptiles to hatch. Bear in mind that the eggs of reptiles' hatch at a colder temperature than that of birds.

In particular, the eggs of lizards' hatch at a temperature of around 20 degrees Celsius or 70 degrees Fahrenheit. Fortunately for the flying reptiles, the daytime temperatures, in early May, at least in this part of the world, frequently reaches that temperature.

Of necessity, in early May, the instinct of the females to reproduce, becomes greater than the instinct to avoid the light. For that reason, the female pterosaurs come out of the nest, choose a hilltop which is reasonably bare, but not too high, build a nest and lay their eggs. They then stay outside the caves and guard the eggs. The animals are able to change their skin colour, to match their surroundings, in much the same way that a chameleon can change its skin colour. In this way, they "blend in", becoming almost invisible.

When the temperature drops during the night, they sit on the eggs. Their bodies act as insulators, so as to ensure that the heat, which the eggs have absorbed during the daytime, stays within the eggs.

I should add that the place where the female builds a nest and lays her eggs, I refer to as the *nesting site* of the pterosaur. I use the term "site", in order to distinguish it from the *nesting ground* of the animal, which is the caves in the mountains. The nesting site varies from year to year, as every predator in North America is constantly looking for those eggs. As well, bare hill tops do not stay bare for long.

Many years ago, one of the Elders in the village, found a rather distinctive egg. He said it "was the size of an ostrich egg, brown, and had a thick skin". It is significant that the eggs of all reptiles have a thick shell. Of course, he had no way of knowing that the egg which he had stumbled upon, was worth a fortune!

As the eggs of pterosaurs are the size of ostrich eggs, that gives us a place to start. It so happens that the eggs of ostriches take forty-two days to hatch. It is reasonable to assume that the eggs of pterosaurs take a similar time to hatch. If that is the case, then the eggs should hatch in mid-June.

Such a hatch places the females upon the horns of a dilemma. *Reptiles do not feed their young!* Once the eggs hatch, then the youngsters are on their own! They have got to fend for themselves! The trouble being that the bare hill tops, while providing heat from the sun, also provides for a breeze. This breeze serves to drive away the bugs. These bugs are the very thing the youngsters need, in order to survive.

I should mention that the newly hatched eggs of birds, are referred to as chicks. By contrast, the newly hatched eggs of pterosaurs are referred to as "flaplings".

As soon as the eggs begin to hatch, another instinct "kicks in". The female picks up the nest, complete with eggs, and carries it to a place where there is a great deal of food for her flaplings. That place is a swamp. The food is mainly bugs, of all varieties, although I am sure they feast mainly upon mosquitoes. Most nourishing! No doubt, they also prey upon small game, such as mice and voles, as well as fish.

Within possibly six weeks, the flaplings are strong enough to fly. This is a fact, as they have been seen flying, at the end of July. Remarkable!

To proceed. As the eggs are the size of ostrich eggs, it stands to reason that the flaplings are the size of ostrich chicks. Further, as the wings are not fully developed, the freshly hatched flaplings cannot fly, so that they are forced to run around on two legs. The problem is one of eating as much as possible, while avoiding predators. Eat and avoid being eaten! Not an easy task!

Countless predators prey upon these flaplings, including birds of prey. Not a difficult task, as they are almost completely helpless. So how to explain that a few of them survive?

Perhaps we can compare them to crocodiles, also reptiles. As soon as the eggs of crocodiles' hatch, the female picks them up and carries them to the water. Then the female guards the youngsters, as best she can, for a certain time. In this way, the odd youngster survives, to adulthood. Perhaps one in a hundred.

It is reasonable to assume that the pterosaurs also stay close to their young, protecting them, as best they can. Then, as soon as the few surviving flaplings are able to fly, to guide them to the nesting grounds, in the mountains. As yet, we have no way of knowing.

It is significant that the Dene refer to these flaplings, in their own language, as "Little People". Perfectly understandable, as they are small, and run around on two legs. They have also been seen in various other parts of the world but are known by other names. The most common name is that of "leprechaun".

Bear in mind that these flaplings, "leprechauns", "little people", will be seen only in early summer, and only around swamps.

As I have a twisted sense of humour, I once made the mistake of mentioning to a couple of girls, that as they objected to being referred to as "chicks", they should count their blessings. They should perhaps be grateful that they are not called flaplings, or leprechauns! I even performed my finest W.C. Fields imitation, of "my little leprechaun".

They responded by telling what they thought of me and my little joke! The language they used left no room for any misunderstanding! Strangely enough, not everyone appreciates my sense of humour!

Pterosaurs Nocturnal

On a more serious note, other questions came to mind. This called for a little research, which in turn provided me with a few answers. It bothered me that such a huge predator, should choose to be nocturnal.

It is the scientific opinion that the pterosaurs evolved around two hundred forty million years ago, on a huge landmass, which they refer to as "pangea". Those same scientists refer to this landmass as a "super continent", which it was not. It amounted to the grouping together of all seven continents of the world, so that it was not one continent, but seven continents, and there was nothing "super" about it. For that reason, I refer to it as the "world landmass" of pangea.

It was on this world land mass of pangea, that the pterosaurs evolved. As the continents were all connected, the pterosaurs spread to all seven continents. This to say that they spread around the world.

At least, they spread around the seven continents. They may or may not have spread to various islands, such as Hawaii or Tahiti. That remains to be seen.

Over a period of many millions of years, the world land mass of pangea split up, and the continents went their separate ways. Each continent carried the pterosaurs with them. As the continent of Antarctic drifted ever closer to the south pole, it gradually became too cold for the eggs of these reptiles to hatch. For that reason, the pterosaurs died out in the Antarctic. *The pterosaurs still exist on the other six continents!*

I can only stress that these pterosaurs still exist on Asia, Africa, Europe, Australia, North America and South America. That is a fact. It is also a fact that they are predators. They hunt in darkness, in open areas. With that huge wingspan, they do not hunt in heavy timber.

All carnivores have a keen sense of smell, as do most predators, and these animals are no exception. Further, all predators are unpredictable, but are far more likely to attack, when they smell blood. For that reason, they frequently prey upon girls of childbearing age. They also prey upon children, at every opportunity, as children are easy to pick up and carry away.

Without doubt, at the time these flying reptiles first evolved, the males had large beaks and head crests. These served as "decorations", adornments to impress the females, during mating season. It is also possible that these long beaks served some practical function, such as catching fish. They almost certainly were active during the daytime, as they had no competition, aside from each other. They had no reason to avoid the light of day.

Competition From Flying Birds

That all changed, and most dramatically, when the first birds "took to the air". Some scientists are of the opinion that archaeopteryx was the first flying bird. Others are not so sure. All are of the opinion that certain species of birds evolved the ability to fly, perhaps one hundred fifty million years ago, or 150 MYA, to use the scientific shorthand. The precise details are in dispute. There can be no question that a great many species of bird began to fly at around that time. Equally without doubt, some of those species of birds were "raptors", birds of prey.

This brings us to a term which has caused a great deal of confusion, and that term is "dinosaur". The word literally means "terrible lizard", and was first coined many years ago. It has since become a household name and is deeply entrenched. That is most unfortunate, as the name is completely inaccurate. For that reason, I try to avoid it, whenever possible. It is not always possible.

The misunderstanding began about two hundred years ago. At that time, the scientists of the day discovered the fossilized remains of huge animals. They were at a loss to explain the existence of these animals, as they could find no reference to them in the bible. At that time, the bible was their only reference source. This was in the days before Darwin. So, they did the best they could, and called these extinct animals dinosaurs.

I have made my position quite clear in other writings. I maintain that all of the animals which are commonly referred to as dinosaurs, had one thing in common. They laid eggs. The egg layers which had feathers, were birds. The egg layers without feathers, were reptiles.

Further, certain species died out, which happens, while other species evolved, and gave rise to new species, which also happens. Among warm blooded animals, such as birds, this happens on a rather regular basis. It is less frequent among cold blooded animals, reptiles.

This is to stress the fact that there was *no mass extinction of dinosaurs!* That whole "theory" is a mere myth! A scientific myth, but a myth nonetheless!

This in no way changes the fact that countless scientists believe in this "mass extinction", and argue passionately concerning the cause of that extinction. Some argue that dinosaurs were limited to warm blooded animals, while others maintain that flying reptiles were also dinosaurs. Certain individuals even argue that those warm-blooded dinosaurs included flying reptiles! Others maintain that the mass extinction of dinosaurs included that of five orders of reptiles! At the same time, they deny that there has ever been a mass extinction of reptiles! No wonder the public is confused!

To return to the time, many years ago, when raptors first challenged the flying reptiles for control of the airways. The war was immediate and brutal. No quarter! Strictly a matter of survival of the fittest. There was no middle ground. The results were clear for all to see.

As anyone who has ever eaten a traditional Christmas dinner can testify, we feast on roast turkey, a bird, and not roast pterosaur, a reptile. Clearly, the raptors won the war for control of the sky, or at least during the daytime.

How is it that the raptors were able to drive the flying reptiles out of the daytime sky, into the relative safety of the darkness? Was their flying ability superior to that of reptiles?

To answer that question, we must examine the flying skills of birds, reptiles and bats, which are mammals.

We know that in order for a body to become airborne, there must be a lifting force on the wings, that offsets the downward force of gravity. That lifting force is created when a wing is curved on the top and flat on the bottom. As the animal flaps its wings, the air that travels over the curves must travel faster than the air below the wings. The difference in air speed creates an area of low pressure above the wings. The air from below the wings pushes up on the wings, which creates lift. Of course, the wings of birds have a curve on the leading edge. The flapping of the wings creates lift, and the animal becomes airborne.

Modern aircraft operate on the same principle. Flaps on the leading-edge combat turbulence, which can be a problem when air speed is low and the aircraft is trying to land.

We can first examine the flight of bats. These mammals are very efficient fliers, and for good reason. The hind quarters of a bat, commonly referred to as its "legs", are attached to its wings. The flapping of the wings are powered not only by the front quarters, called the "arms", but also by the legs. When the bat flaps its wings, the legs move up and down with the arms. The result is a very strong wing stroke, which results in greater lift. The tracks left by the bat are also distinctive, as they walk on the front paws as well as the hind paws, with the hind paws splayed out, far to the side.

The recently discovered fossilized trackways of the pterosaurs show that the animal walks in a manner similar to that of a bat, sometimes on two legs and sometimes on all fours, with the hind paws spread out, clear evidence that the hind paws are attached to the wings. All four limbs of the pterosaurs are involved in the flapping of the wings. Clearly, the pterosaur is a most efficient flier, far more so than I expected. Almost all the muscles of the body are involved in the flapping of the wings, which is to say flight.

It is also a fact that this animal has a special plate of light weight bone that strengthens the torso and shoulders, thus eliminating most of the muscle groups that do not contribute to flight. This gives it an advantage. The more light weight the flying animal, the easier it is for the animal to become airborne and stay airborne.

As well, the pterosaur has flaps on the leading edge of the wings to combat turbulence, thus making it more stable, just as modern aircraft have flaps.

Even this is not the only flying advantage of the pterosaur. The shoulders of the animal are a marvel of aerodynamic engineering. It is very likely that this animal has an extra pivot joint between the upper end of the shoulder blade and the bony plate that stiffens the torso, which allows the shoulder to swivel. This permits the shoulder bone to swing not only up and back, but also down and forward. Such movements dramatically increase the power of the down stroke of the wing.

In addition, as cold-blooded animals, reptiles need to consume far less food than birds, which are warm blooded animals. These are all to the advantage of flying reptiles.

We can contrast these characteristics of reptiles, to those of birds, starting with the flight feathers. Such feathers are attached to the wings and only to the wings. Only the front quarters of birds are involved in flapping the wings. The hind legs of birds are dead weight, when the animal is flying.

On the other hand, as warm-blooded animals, birds do not require the heat from the sun to warm up and become active. At the first crack of daylight, birds, including raptors, took to the air and began to hunt, as they do to this day!

By contrast, pterosaurs, as flying reptiles, had to wait for the heat from the sun, to warm them up, before they could become airborne. The amount of time required to warm up the reptile varied, depending upon the size of the reptile, the night temperatures, the weather conditions, the time of year and numerous other variables.

The brief interval, between dawn and the time that the heat from the sun warmed up the flying reptiles, was critical. Although that interval may not have been long, it was long enough for the predatory birds, the raptors, to decimate the flying reptiles, the pterosaurs. Even though the cold-blooded reptiles were far more efficient fliers and required far less food, pound for pound, the raptors still out competed them for mastery of the daytime skies.

It is very likely that over a period of tens of millions of years, various species of raptor evolved, and gradually drove all species of flying reptiles, pterosaurs, either to extinction, or to the relative safety of darkness. This is another way of saying that the flying reptiles became nocturnal, and this happened around the world.

This is not to say that the raptors have won the war, and there is no more competition between birds of prey and flying reptiles. It just means that the battle ground has changed, from the daytime skies, to the nighttime skies. Raptors such as owls and night hawks, hunt in the darkness, as do bats, in competition with flying reptiles, pterosaurs.

Pterosaurs and UFO's

As the pterosaurs gradually adapted to a life of living in darkness, they were forced to make a few adjustments. In particular, during mating season, the females were no longer deeply impressed by the long beaks and head crests of the males, if only because they could not see them. The males had to find another way to display, and they did. The males evolved the ability to glow.

As for those who find this quite remarkable, consider the fact that "fireflies" have evolved the ability to glow. These insects, otherwise known as "glow worms" or "lightning bugs", are really beetles. Light production in these beetles is due to a chemical reaction called bioluminescence. The beetle has a special light emitting organ in its lower abdomen, just as there is no doubt a similar light emitting organ in the pterosaur.

In the case of the firefly, there is an enzyme, luciferase, which acts on the luciferin in the presence of magnesium ions, ATP, and oxygen, to produce light. The light is in the visible spectrum of 520 to 680 nanometers, so that we see the light as yellow, green or pale red.

No doubt, a similar chemical reaction occurs in the pterosaur, which gives off a glow from the torso, and is clearly visible from the ground.

As yet, we do not know the precise wavelength of light, which is to say the colour, of the light produced. We do know that each species glows in a distinctive colour. These colours are as distinctive as the plumage on birds and will soon be used to help identify the species of pterosaur.

It is to be hoped that many readers will be inspired to look into the chemistry involved in the production of such bioluminescent light.

It is very likely that the males acquire the ability to glow, only at the onset of sexual maturity. As yet, we have no way of knowing if the females also have the ability to glow.

The reason I say this, is that there are times when evolution will take one characteristic and use it for more than one purpose. Herds of caribou come to mind. Among such animals, both males and females grow antlers. The males grow antlers in the spring and summer, in order to display in the fall, and then drop the antlers in the winter. But then the females grow antlers in the fall and winter, in order to protect their calves, which are born in the spring, and then drop the antlers in the summer.

In much the same manner, among pterosaurs, the males use the ability to glow, in mating season, as a means of display. But then those same animals also use the glow for hunting purposes! They will fly around a huge swamp, glowing brightly. The bright light attracts clouds of insects, and those insects attract bats. Then the flying reptiles attack the flying mammals. As yet, we have no way of knowing if the females also glow, under these circumstances.

We do know that many people find these lights to be quite entertaining, and in certain communities, are used as a tourist attraction.

We also know that this glow, from the torso of the pterosaurs, is the basis of a great many Unidentified Flying Objects, or UFO's.

As I have now identified many of these flying lights, it is no longer correct to refer to them as UFO's but as IFO's, Identified Flying Objects. Of course, that is not about to happen, but hopefully, it will appease many of my most vocal critics, those who have been complaining that I have not been looking for UFO's.

God forbid that I should be accused of being a UFOlogist! For many years, it has been my opinion that such people are not entirely sane, to put it politely. More accurately, I have long maintained that they are completely out of their minds! I can only hope that they will adopt a more scientific approach to their research.

Death and Mutilation of Livestock

These lights in the night sky, so called UFO's, have been associated with the death and mutilation of livestock. A popular television show has recently stumbled upon the fact that in the mountains around the world, horses and cattle which are left outside, after sundown, are being discovered in the morning, "dead and mutilated". The fact that there is no blood around the wound sites has everyone puzzled.

The writers of the television show have not stated it quite that clearly. But then, they have no desire to be accurate. They are focused on entertaining, not educating. For that reason, they have come up with some hare-brained theory to explain this phenomenon. The "explanation", as put forward by the announcer, is that "aliens from a distant world have travelled to planet earth", presumably at "warp speed", -whatever that is-, have "beamed" these animals up to their space craft, drawn out their blood and mutilated them, and then "beamed" them back down to planet earth.

Further, as "proof" of the existence of these spacecrafts, the commentator has mentioned the existence of these lights in the night sky. He maintains that the "motion of these lights does not match the motion of any mechanical aircraft, either plane or helicopter".

At least that part is true! It is not the motion of a machine, but that of a flying reptile! It is entirely possible that the announcer actually believes the nonsense he is spitting out!

Perhaps his days as an actor on a popular television show, in which "beaming" was commonplace, has caused him to lose touch with reality!

In all fairness to the writers, they have drawn attention to the fact that livestock is being killed and mutilated, in the mountains, "around the world". They did not state it quite that clearly, if only because they were hindered by their mysticism. Or perhaps their only concern was that of entertaining.

No doubt, the "lights in the night sky", the "UFO's", are seen only in mating season, in the spring. Another little detail that the writers neglected to mention! That in no way changes the fact that livestock, which is left outside after sundown, is being found in the morning, dead and mutilated.

The owners of these animals are completely puzzled, as they have clearly not been killed by humans. There is no indication of bullet holes, puncture wounds or signs of strangulation. Nor have they been killed by predators, such as bears, wolves, grizzlies or cougars. The fact that there is no blood around the wounds, leaves no room for any misunderstanding on that point!

This lack of blood around the wound sites is of critical importance. It does not mean that the blood is missing! It tells us that the animal was dead when it was mutilated. The question then becomes, what killed the animal?

We can rule out blood loss, so that leaves poison. What kind of poison? It certainly did not eat the poison, so it must have inhaled the poison.

The livestock is clearly being killed by a pterosaur.

The reason I say this, is because this animal is famous for releasing a so called "cloud of smoke". Except that this "cloud of smoke" is not smoke at all. It is toxic. Poison gas. Strong enough to kill a horse! We know this for a fact, because that is precisely what it does! And not just horses! Also, cattle! Further, they prefer to prey upon the males, because the first thing they go for is the genitals. Also, the flesh from the face, as well as the large intestine.

Incidentally, this "smoke" which the animal releases, is the basis for the legend of the "fire breathing dragon".

The pterosaur is not strong enough to rip apart the carcass, so it has to settle for the morsels it can gather.

The standard response of the farmer is to call the police. As he is well aware that the animal was not killed by a predator, or at least not by any predator of which he is familiar, he just naturally assumes that it was killed by humans. Such an assumption is natural but mistaken.

As the police respond, they too are puzzled, as it was clearly not killed by any human. They note the lack of bullet wounds, footprints, tire tracks or blood. Clearly not a police matter, as no crime has been committed. For that reason, they leave.

This does not satisfy the farmer, so frequently in America, the FBI is notified. They in turn refuse to respond, as they have seen this so often. This response may be correct but does not solve the problem.

Technically, this is a matter for the game warden. Not that the game warden is ever notified! Even if he is, this is beyond his level of training.

The scientists should be involved, but they are careful to avoid the subject. The last thing they want to do, is prove the existence of the pterosaurs! And proving that is not difficult!

Three things have to be done. The first two are quite simple and easy, while the third is more challenging.

The first thing that has to be done, is a simple matter of drawing out a sample of blood, from the carcass, and sending it to the lab. As long as the blood is drawn out within twenty-four hours, after the death of the animal, the lab should be able to determine the poison gas that was used to kill the animal. I am certainly curious.

The second thing is also quite simple. The wounds should be swabbed, and the swabs sent to the same labs, for a DNA test. The lab will report that the DNA is that of a reptile, one which is not known to science! It stands to reason that this will prove that these animals were killed by poison gas, released by a reptile. What reptile?

The answer to that question involves getting government permits, with the goal of opening up an old logging road, close to the village in which I live. It is called the Ten Thousand Road. The entrance to the caves, which are the nesting ground of the pterosaurs, open up onto that old Road. Fortunately, it is once again being opened. The first of two bridges is in place.

Now it is up to the Forest Service to put in place the second bridge and further open up that Road. Those caves are just beyond the second creek.

Then it is a simple matter of placing trail cameras at the openings to those caves, as the animals come out of those caves after sundown, but before dark, and return to those caves before sunrise, but in the daylight. It is also a fact that, after they come out of those caves, they then climb to the top of the mountain and open their wings. That mountain top is also a good place for a trail camera.

This calls for a little explanation. Thirty years ago, we were logging up that Road. My relatives pointed those caves out to me and told me that they lead directly to hell. They added that after sundown, Satan opens the gates of hell, and turns loose his demons, in the form of birds.

As I respect the personal beliefs of all people, I made the mistake of taking them "at their word". I thought it was just a personal belief. Yet these beliefs are based on facts.

Of course, after the area was logged out, it was deactivated, and I missed my opportunity to set up trail cameras.

Disappearance of Girls of Childbearing Age From Open Areas

It is vitally important to prove the existence of this animal. The members of the public have got to be made aware of the fact that it exists. It is a predator, it hunts in darkness, in open areas. It has a keen sense of smell, and is far more likely to attack, when it smells blood. This helps to explain the disappearance of so many girls of childbearing age!

The Highway from Prince Rupert, on the West Coast, to the city of Edmonton, in the central interior of the province of Alberta, is known as the

Highway of Tears. It has earned that name, due to the disappearance of so many people, over the years, from this stretch of the Highway.

It is not by chance that this Highway runs directly through the mountains! Nor is it by chance that most of the people who have disappeared, have been girls of childbearing age!

The sad fact is that families break up, and at that time, all too often, young girls leave home. In the cities and towns, they generally jump on a bus. Human predators are well aware of this, and consistently hunt for those young girls, at bus depots. By contrast, the girls of families who live in isolated areas, including many Reserves, tend to go hitch hiking. These girls are then frequently preyed upon by pterosaurs. They generally attack because they smell the blood!

Of course, it is not just on highways that people are attacked by these flying reptiles. They hunt in all open areas, in the darkness.

The areas have to be open, because the animal has a huge wingspan, and the area has to be dark, because the animal tends to avoid the light. On the other hand, there are indications that the animal is becoming "braver", more familiar with artificial lights, so that such lights do not offer full protection, as was once the case.

I have been asked what to do, if caught outside, after sundown, and attacked by the pterosaur. Find shelter! With that huge wingspan, they are forced to hunt in open areas. If no shelter is available, hit the ground! They hunt us by wrapping their paws around our shoulders and sinking their claws in! Do not make it easy for them!

If we are flat on the ground, then that cannot be done! By all means, fight them! They *may* kill you, but if you do not fight them, they *will* kill you!

I have also been asked to clarify, the "sound of flapping wings", of that animal. To say that it "sounds like dry hide" is not terribly helpful, unless you are familiar with the sound of dry hide! Most people are not!

With that in mind, perhaps it would be best to allow a girl who was attacked by a pterosaur, to describe the event, in her own words: "It was after sundown, but not yet full dark when we first heard a whooshing sound. Really strange. Whoosh, whoosh, whoosh. Then we looked up and saw it. It attacked us and we chased it away, so we followed it".

For what it is worth, the flapping of the wings makes a distinctive "whooshing" sound.

As for those who think that the paws could not possibly be big enough to wrap around our shoulders, think again. Those who have seen the tracks that these animals leave in fresh fallen snow, can testify to that. They assure me that the paws are possibly half a meter long or 18 to 20 inches, and 6 to 8 centimetres or 2 to 3 inches wide. As they phrase it, "real long and real skinny". It is not by coincidence that these descriptions match the fossilized trackways of the pterosaurs!

After the animal comes out of the caves, in the darkness it generally selects an evergreen tree, and with its paws, pushes over the top stem of the tree. Then, with the stem of the tree horizontal, or "flat", to state it in common terms, the animal sits on that perch.

These trees are usually located at the edge of a clearing, as the animal hunts in open areas. These clearings may or may not be on the edge of a highway or road. The animal also hunts near campgrounds, playgrounds, school yards, and in fact near any open area, and that includes the backyards of people. They tend to return to the same tree, night after night. As a result of this, the top stem of those trees tends to grow on an angle, or a "slant", as is commonly stated. To identify these trees is to identify the hunting grounds of the pterosaur.

As it is only the top 4 to 5 metres or 12 to 15 feet of the tree that is bent over in this manner, that provides us with a "ballpark figure", as to the weight of the animal. I estimate the animal weighs no more than 10 to 15 kilos, or 25 to 35 pounds. I refer to this as an example of using "common sense", and it is one of the methods I recommend.

I do not regard this as an alternative to the scientific method, but as part of the scientific method.

I should add the fact that as the animal prefers to sit on the top of evergreen trees, this helps to explain the reason that experienced woodsmen prefer to build their cabins within clearings. It is partly to avoid the chance of a strong wind blowing a tree down, onto the cabin, but also because such woodsmen are aware of the existence of this animal. They are also aware that it likes to sit on the top stems of trees.

In the winter months especially, the woodsmen have to come out of the cabin after dark, if only to go to the back house, or wood pile. If a tall tree is close by, they have no way of knowing if a "Devil Bird" is sitting there, just waiting for them to step outside.

These animals have truly adapted to a nocturnal lifestyle. They come out of those caves only after sundown (aside from the females, which emerge in order to build a nest and lay eggs), at a time which is generally cooler than the daytime temperatures. In fact, these reptiles are able to hunt in temperatures as cold as twenty degrees below zero Celsius, or zero degrees Fahrenheit.

These reptiles are able to do this, but only because they are very large. Only the largest of reptiles are able to withstand such cold, and only for short periods of time, due to a process called gigantothermia. It is precisely the same process that allows another species of reptile, the leatherback turtle, to survive in the very cold environment of the North Atlantic.

While outside the cave, the pterosaur loses little heat, partly because a large body size leads to a small surface to volume ratio, so that the heat exchange volume remains low. Consequently, the core body temperature is slow to change, while a spherical body and a layer of fat helps a great deal.

Local people know the animal is close by and on the hunt, because they can recognize the sounds that it makes. It is able to precisely imitate the sounds it hears. Around here, that includes the sounds of "dogs barking, babies crying and women screaming". Without doubt, they absolutely hear the sound of women screaming. In fact, each time they attack women, they hear the screams! In addition, the girls tell me that it can even imitate the sound of the human voice! No wonder it is frequently confused with a Demon!

One of the girls in this village was recently attacked, by this animal. She had a campfire going behind the house of her father. An open area! In the darkness, the animal attacked, probably because it smelled blood. At the time the animal struck, it releases a very loud whistling sound! She was not physically injured, by the Grace of God!

Terrified, she ran into the house! Yet her ears were ringing! The whistling sound was that loud! This explains the origin of the name "Thunder Bird".

This same girl called me, very upset. She told me that she mentioned this attack to her friends, and they laughed at her! They said she must be on drugs! First, she was very nearly killed, and then her friends laughed at her! Adding insult to injury!

My response: "Welcome to my world!"

Incidentally, closer to town, they hear the sounds of car horns and train whistles, so no doubt they are able to imitate those sounds also. The animal is also reasonably intelligent, at least for a reptile.

I was surprised to find that they have learned to associate the sound of gun shots, with food. The local boys found that out, at the time they were "spring trapping", several years ago.

At that time of year, March and April, the fur of the land-dwelling animals is no longer valuable, as the warm daytime temperatures adversely affects the winter coat of the animal. By contrast, the fur of water dwelling animals, such as beaver, mink and muskrat, remain in prime condition. For that reason, the local trappers hunt them, focusing mainly on beaver.

Their method is to set traps for them, as well as sometimes shooting them, as they come out of the water. The boys learned that on the days they shot beaver, they could expect "company" after sundown. By contrast, on the days they merely trapped beaver, no shooting, no "Devil Bird" visit after sundown. The sound of gun fire is a "dinner bell" to this animal!

This brings us to Missing and Murdered Indigenous Women, MMIW. This is a painful subject, thought to be a national disgrace, as indeed it is. Which is not to say that the title is entirely accurate, as it is not only Indigenous women who are disappearing. Members of other ethnic groups are also disappearing, as well as children.

As previously mentioned, most of these people are females of childbearing age, which the animal attacks, because it smells blood.

Then too, children are easier to carry away. I mention this again, as it is so important. People must be warned! No doubt, some of these disappearances are due to human predators. Yet a great many of those that disappear from open areas, in the darkness, have been killed by pterosaurs.

This brings me to a very sensitive subject, one which I have previously avoided, out of respect for the family. With that in mind, my most sincere condolences to the family of Madison Scott. I cannot imagine the mental anguish you are suffering. But perhaps her disappearance and death can count for something.

Having said that, the fact is that Madison disappeared from a campground near Vanderhoof, twelve years ago. I suspect she was picked up and carried away by a pterosaur. If that was the case, it is very likely that her friends heard a very loud whistling sound, at the time the animal struck. Also, very likely that her shoes were left behind, as she was being dragged away.

Recently, her remains were located. It is entirely possible that the animal lost its grip. If that was indeed the case, then the bones would have fractures that are characteristic of a fall from a height. A forensic anthropologist can determine such fractures. As well, we can expect to see fractures in, the upper thoracic region, in that certain ribs, as well as the clavicle and scapula, can be fractured. As well, in case of a severed subclavian artery, we can expect to see blood-stained bones.

The police are not releasing any details, so for now, it remains mere speculation. I mention this only because it is so important.

Across the country, various police departments are under extreme pressure. It is assumed that there are possibly several "serial killers", men who are responsible for the killing of these girls. There is even an RCMP "Highway of Tears" task force.

The police cannot be faulted for failing to prove the existence of this animal. It is up to the scientists to prove they exist. I stress, it is the *duty of scientists!* The scientists are negligent in doing *their duty!*

It may be objected that I am being too harsh with the scientists. Some people have even gone so far as to say that the scientists may not be aware of the existence of the pterosaurs. To such kind, sensitive, tender-hearted souls, I can only respond: *How can they not be aware of their existence?* They are the size of small planes! Even the most passionate, starry-eyed optimists, should agree that it is the duty of scientists to explain that which puzzles people.

Countless people are puzzled by UFO's, including " cigar shaped UFO's", and "Flying Saucers". As well, the "death and mutilation" of livestock, on farms, is widely known. Yet the scientists remain silent!

As the scientists refuse to perform their duty, it is up to common people, to do their duty for them. Granted, that is not fair. That is not the way it should be. It is what it is!

On the other hand, it is not terribly difficult. A simple blood test and a DNA swab. Acquiring permits to open up an old logging road is more difficult. Then there is the matter of raising money to hire the machines. But it can be done.

To readers of this book, I can only say: I am counting on you!

The Role of Capitalism In Science

The fact of the matter is that it is the duty of the scientists to prove the existence of these huge animals, and not merely the pterosaurs. Yet they are also supremely well aware that such an action would almost certainly cost them their careers!

The class of people who are currently in charge, the monopoly capitalists, the billionaires, are determined that nothing will change. Any scientist who challenges any of the currently accepted scientific theories, such as the mass extinction of dinosaurs, is subject to immediate dismissal. The extinction of pterosaurs, sixty-five million years ago, is one of those "sacred scientific theories". Not to be challenged!

The billionaires cannot threaten me with career suicide, as it is a little late! I did that to myself many years ago. I no longer have a career!

With that in mind, I will stress the fact that the animal responsible for this butchery is the pterosaur, an animal thought to be extinct for many millions of years. It most certainly is not extinct! In fact, it is alive and well! It preys upon animals as well as people and is supremely dangerous.

Cattle Killed and Mutilated In the State of Oregon

Of late, the press has been reporting that a great many livestock, both cattle and horses, are being killed and mutilated, in the state of Oregon. If nothing else, this is a somewhat welcome change from the reports of the political gong show in Washington. Aside from that, the speculation as to the cause of the death of these animals is nothing less than an embarrassment.

In the eastern part of the state, there are very few ranches, but those which do exist are huge, covering many thousands of acres. The cattle are allowed to range freely. It is on these huge estates that the cattle are being discovered, dead and mutilated. There is no shortage of theories to explain these strange events.

As well, it is significant that the ranch owners are careful to provide a ratio of one bull for every twenty-five cows, approximately. There are far more cows than bulls! Yet most of the animals which are being killed, are bulls!

Each time this happens; the local police are called in to investigate. In one case, the police report stated, there was "no indication it had been shot, attacked by predators or eaten poisonous plants. The animals sex organs and tongue had been removed. All the blood was gone".

Here we have an example of an incomplete police investigation.

In fact, just because there was no blood around the wound sites, does not mean that the "blood was gone". The blood was precisely where it was supposed to be, inside the carcass. If the police had conducted a proper investigation, that would have been confirmed.

In another case, that of four bulls being discovered dead, the local press reported that: "There were no tracks around the carcasses. Ranch management and law enforcement suspect that someone killed the bulls. Ranch hands have been advised to travel in pairs and to go armed".

As there were no tracks or signs of human activity around the carcasses, no bullet holes or puncture marks, and no signs of strangulation, the conclusion is illogical. The bulls were clearly not killed by people.

It is further reported that, in almost all cases, the genitals of the bulls have been removed, with "surgical precision", so that it is speculated that the bulls were "killed to get the organs of the free ranging bulls". It is further speculated that the bulls were "darted with tranquilizers that knocked them out. While some people acted as lookouts, others bled the animals by inserting a large gauge needle into the tongue and into an artery, then removing the organs after the heart quit beating."

Once again, we have an incomplete investigation, in that it is assumed that the blood is missing. Also, there was no sign of human activity, a fact which was conveniently ignored. As the country is supremely dusty, any tracks are immediately apparent. That rules out human involvement. As well, they are correct when they say that "It is not bears, wolves, cougars or poisonous plants. Nor were the animals shot.... not one drop of blood". Yet they are determined that people were responsible for this!

Strangely enough, these same people are the first to admit that it is simply impossible to move around that "dusty country" without leaving tracks. On the one hand, their investigation rules out human activity, yet on the other hand, they still blame people!

In the case of the death of one animal, according to a deputy conducting the investigation: "No dart puncture, no bullets, no strangulation marks, no rope burns, no tire tracks, no signs of poison". From that rather thorough investigation, the same deputy reached a rather strange conclusion: "definitely foul play involved in this animals' death". Once again, we have incorrect conclusions derived from incontrovertible facts.

It is perhaps significant that this is quite common. Even highly trained professionals refuse to rule out the impossible and face the one and only alternative! That solitary alternative may be extremely unlikely, but not impossible! As these animals were not killed by humans, and did not eat poison, they must have been killed with poison gas!

Then there is the idea of "people attacking the animals to cause financial harm to the ranchers". We could point out that there are easier ways of causing "financial harm", but even that brainstorm is better than the more popular belief, that "the animals are being levitated into a spaceship, mutilated, and then dropped back to the ground".

It is to the credit of one of the ranchers that he joked about this. He said he was "flattered that aliens from distant galaxies would travel all the way to planet earth, just to kill and mutilate my cattle." He is one of the few ranchers to still maintain a sense of humour.

As local law enforcement is stumped, the FBI has been called in. They in turn report that a similar situation took place in the 1970's. As they put it, at that time, "in the West and Midwest, thousands of cattle were killed and mutilated, from Minnesota to New Mexico. There have been sporadic cases since then". The investigation of that august agency concluded that "there is no indication that anything other than common predators were responsible."

As the investigations have proven conclusively that the animals are not being killed by bears, wolves or cougars, it is clear that the FBI considers some other "common predator" to be responsible. They neglect to say just what predator that may be!

Here we have a very clear-cut example of well meaning, honest, professional, highly respected people, including members of various law enforcement agencies, stating the problem clearly, investigating carefully, and then coming to conclusions. The trouble is that the conclusions they reach are in complete contradiction to the facts! This can only be described as shoddy work.

All are correct when they note that there is no blood around the injury sites. They are also correct when they note no sign of human interference, such as "bullet wounds, puncture wounds, strangulation marks, rope burns or footprints". Also, no sign of predation from "common predators". From this they conclude that the "blood was drained from the animal", or "foul play was involved", or as the FBI stated, "common predators were responsible". All are mistaken, although the FBI is closest to the truth.

Perhaps someone should tell these Dick Tracy sorts, these ace detectives, that which all school children know. Once an animal dies, the heart quits beating. There is no blood around the wound sites, because the animal was dead when it was mutilated! Dead animals do not bleed! They correctly point out that there is no indication of the animal being killed by a human, of "foul play", because there was no foul play! The police have done a fine job of ruling out the impossible but failed to take the crucial next step of facing the one and only alternative! The animals are being killed by poison gas!

There is only one animal which can release poison gas, strong enough to kill cattle. This poison gas, commonly referred to as a "cloud of smoke", is not smoke at all, but supremely toxic. The fact that it uses this poison to kill cattle is proof of that. Then it consumes the parts of the animal it can easily gather, including the tongue and genitals. It also frequently rips out the large intestine, a very nourishing part of the anatomy. It is strange that the press does not report this, perhaps because they are concerned with the sentiments of the readers. Or perhaps the animal has changed its behaviour.

Of course, the animal to which I am referring is the pterosaur, commonly called the pterodactyl. As it is nocturnal, it is rarely seen. It also flies, so that it is very light, with huge paws, to that the tracks it leaves are very difficult to spot. It has also clearly developed a taste for beef, at least in Oregon. As long as livestock is left outside after sunset, in areas close to mountains, then those animals will continue to be preyed upon by this prehistoric monster. I regret I do not have better news.

To the law enforcement officials, I can only say that there is no law against wild animals killing livestock. In other words, no "foul play" involved. To the FBI, I can say that the pterosaur is hardly a "common predator", but it is certainly a predator.

Proving the Existence of the Pterosaur

That brings us to the not so little matter of proving the existence of this "not so common predator". May I suggest taking a sample of the blood from the carcass and sending it to a laboratory. As long as the blood is extracted from the animal within 24 hours of the death of the animal, then it is very likely the lab can determine the poison used to kill the animal. At the same time, swab around the wound sites and send the swab to a lab, for a DNA analysis. No doubt the test result will confirm that the wound was made by a reptile, one which is not known to science. Equally without doubt, those test results will be challenged, so it may be best to send the samples to three independent labs. This is my idea of a thorough investigation.

These test results will confirm that the livestock is being killed by a blast of poison gas and consumed by a reptile. Law enforcement personnel can take comfort in knowing that the ball will then be taken from their court and placed where it belongs, in the court of the scientists. The death and mutilation of this livestock is not a crime. It is a problem to be faced by the scientists. It is high time they did their job.

With that in mind, may I suggest opening up the Ten Thousand Road. The opening to the caves, which are the nesting ground of the pterosaurs, open up onto that Road. It is currently opened up to 12 Kilometre, at which point a second bridge must be put into place. The caves are just beyond that creek, at around 15 Kilometre.

PART 3

SEPARATE SPECIES OF HUMAN

Sasquatch or Bigfoot

The idea that our species, that of homo sapiens, is unique, has a certain appeal, especially to common people. It must also have a certain appeal to the scientists, as it is deeply entrenched in the scientific literature. They would have us believe that even though various species of humans evolved, we are the one and only species of human to still survive.

In fact, all scientists are agreed that of all the species of human that have ever evolved, we are the one and only species of human still living. They are mistaken! All reports, many of which are highly reliable, of "giant naked hairy humans", are carefully ignored. As well, countless plaster casts of footprints are dismissed as fakes.

The fact of the matter is that a species of ape, known as Gigantopithecus, managed to achieve bipedal locomotion, which is to say that they started to walk on two legs, and then, with their forelimbs free for labour, evolved the opposable thumb. They can touch their finger tips with their thumbs! Bipedal apes, complete with the opposable thumb, are the very definition of human! At least, that is my briefest possible definition of human.

As Engels pointed out, as the paws of the ape were transformed into hands, complete with the opposable thumb, corresponding changes also took place within the rest of the anatomy. These changes we cannot as yet explain, but only describe, and then "only in general terms". Taken together, they make us human.

I maintain that gigantopithecus is a separate species of human, currently walking the earth, here in North America. They live among us! That species is commonly referred to as Sasquatch or Bigfoot. I refer to them as Giants, as they are so huge.

The Dene refer to them as "Stink People", in their language, a title which is less than flattering. Yet the term is not meant in a derogatory sense, as the Dene have the utmost respect for those people.

Instead, it is a reference to the "putrid smell", the "terrible odour", of which they are justifiably famous. This has led to the mistaken belief that these people "smell terrible". They do not.

This stench is a further indication of their humanity! They are a sensitive lot, and do not take well to insults. For that reason, at the time we insult them, they respond by releasing a noxious blast of poison gas, from their rectal orifice. One good turn deserves another!

This is commonly referred to as "tooting", or "farting", and is considered to be most impolite. As it is. Yet the fact that they resort to such impolite behaviour, is further proof of the humanity of those people. It is the universal, non verbal display of contempt!

The scientists believe that this species of ape evolved many years ago in Asia. As it stood ten feet tall or three meters, and weighed twelve hundred pounds or five hundred kilos, it was by far the largest ape ever to walk the earth. Twice the size of a gorilla! But then, the scientists maintain that around one hundred thousand years ago, this huge ape, gigantopithecus, dropped dead, for reasons which no one can explain.

The scientists are mistaken. Gigantopithecus did not go extinct! They first became bipedal, and then they developed the opposable thumb. In other words, they made the transition, from ape to humans. As humans, they then began to bury their dead. That is the reason no recent remains of these people has been discovered.

Of course, all scientists disagree with me.

That in no way changes the fact that these people, these Giants, live among us, and are seen on a regular basis! As members of a separate species of human, they deserve our utmost respect. They have every right to live their lives as they see fit. Our laws do not apply to them. We have no right to interfere in their way of life. We certainly have no right to hunt them, as so many people are hunting them, as if they were vermin, determined to kill them.

Such a killing can be seen as nothing less than an act of murder, completely unjustified. The fact that many of the people who are hunting them are doing this with good intentions, does not change that fact. Those people think that this is the only way to prove they exist. The end does not justify the means.

I maintain that the Giants are members of a primitive, hunting gathering society, so that they are constantly travelling. To stay for any considerable length of time, in any one area, would quickly result in exhausting the food supply, among other things. For that reason, they are constantly moving, following the herds. As well, the wild plants are harvested in the proper season, in the proper location.

In the spring, they travel north, in order to take advantage of the abundant supply of eggs of birds, as well as the birth of numerous animals, such as deer, moose, elk and caribou. They also harvest the wild berries as they become ripe. In the fall, they travel south, in order to take advantage of the warmer temperatures, as well as different food supplies.

The further south they travel, the more settlements they pass through. The people who live in those communities take note of the tracks they see in the morning. They know full well that something big had placed those tracks the previous night, while overlooking the fact that at the time the tracks were being placed, the dogs were not barking.

This silence of dogs is significant, as dogs are known to bark at all animals known to science. As the dogs are terrified to the point of silence, in the presence of Giants, this should give all of us some idea of the size, strength and sheer terror their presence inspires in dogs. If nothing else, this is reason for giving them our utmost respect.

As many people are hunting them, their lives are being made quite miserable. They are constantly on the run. This constant harassment can result in something more than inconvenience. They are members of a hunting gathering society, so their food supply is completely unstable. It varies according to the season and changes in the weather. They have no emergency food supply to fall back upon. They live constantly on the edge of starvation. Hunger is a constant companion.

Danger of Inbreeding

In addition, with our constant expansion and industrial development, we are destroying their hunting and gathering grounds. This industrial development includes dam building, logging, mining and road building, which makes it difficult for the young males to travel to different bands of people, to whom they are not related, in order to find wives. This is the one and only way to avoid inbreeding, and is the only way to avoid the extinction of the species.

This is not an exaggeration. The current European nobility can testify to the fact that inbreeding can lead to disaster. All of them are related, and their off spring, the heirs to the throne, can at best, charitably be referred to as "simple souls".

Anyone who doubts this has only to face the fact that the queen of England and her husband are closely related. Ever since the birth of their first child, they have no doubt regretted the fact that they "tied the knot". The heir to the throne of England is a constant source of embarrassment to the queen.

For that reason, it is now fashionable for the nobility to insist that their young people marry "commoners". This is not to say that they find us any less contemptible, as they do not. They have merely been forced into this.

The "royal watchers" are now speculating that the queen has already decided that the next king of England will not be her first born male child, but her first born male grand child. That child is not as deeply inbred as his father.

The point of this is to drive home the fact that inbreeding is certain to lead to disaster. We have no way of knowing the population size of the Giants, but we do know that as members of a hunting- gathering society, their numbers are limited by their food supply.

This is another way of saying that there cannot be a great many of them. If their population is reduced to a very low level, or if the males are unable to travel to other groups in order to find wives, then the species is certain to go extinct.

To be responsible for the extinction of another species of human goes beyond genocide. Yet unless we change our ways, we could be guilty of precisely that.

Proper Approach To Giants

For that reason, it is urgent that we prove the existence of these Giants. Many people have been trying to do this, but they are going about it in the wrong way. A different approach is required. Instead of chasing them, attract them! They have no reason to trust us, as we are quite fond of shooting them! It will take a while to convince them that we mean them no harm.

The only members of our species they trust is the Indigenous people. The Giants and the Indigenous people respect each other. They leave each other strictly alone. Live and let live.

On the Reserves, along the west coast of British Columbia, they share the beaches. The Indigenous people enjoy the beach during the day, while the Giants enjoy the beach during the night. It is on the beaches of the Reserves that we can meet them.

The key to this historic meeting lies with the Indigenous Elders. They have about as much reason to trust us, as do the Giants. I can only suggest calling a Reserve on the West Coast and requesting a meeting of the Elders Society. Assuming they agree, then arrange a feast for the Elders. Be sure to bring in food those people love. As well, buy gifts, such as tobacco, Hudson Bay blankets, electric blankets and tanned hides. The management can advise the gifts the people appreciate. Spare no expense, as we are looking at a major scientific break through. Make sure that food and gifts are taken to the Elders who are too feeble to attend the meeting. Respect is essential to the success of this project. With that in mind, offer to hire a translator. It is very likely that all the Elders understand English, but a little show of respect goes a long way.

After the Elders have eaten and received their gifts, explain to them that we are interested only in meeting the Giants. We mean them no harm. On the contrary, there are countless people who are interested in killing them. The one and only way of preventing this is to have laws passed to protect them. The only way this can happen is by first proving they exist. That is where the Elders come in. It cannot be done without the help of the Elders.

With that in mind, mention that the ideal place to make contact with the Giants is on the beaches of the Reserves. This can only happen with the blessings of the Elders and the cooperation of the young Indigenous people. The Elders know the time the Giants are on the beach, and with the help of the young people, gifts can be placed on the beach, close to sundown, for the Giants.

As the Giants are human, there can be no doubt that they love the same things that we love. That includes meat, fish, vegetables and fruit, raw as well as cooked. It also includes cosmetics and mirrors, preferably small metallic mirrors. Decorations of metallic ornaments, tied together with dental floss, is also an idea, as is burlap bags.

As for those who complain that these are merely "beads and trinkets", I can only respond that you are absolutely correct. May I suggest that our goal is to meet these people, to prove they exist. We want to impress upon them that our intentions are honest, and gifts of platinum, gold and silver are not about to do a world of good. They have no use for such items! No doubt all of them want to know what they look like, and most members of our species wear cosmetics and decorations of one sort or another. As the Giants are just as human as we are, it is reasonable to assume that they love the same things we love.

Perhaps it would be best if government officials are not involved in this science project. Such officials tend to have their own agenda, which is frequently at odds with the goal of the project. At best, such officials merely spread confusion. At worst, they sabotage the operation. It is a mistake to expect anything better from such people.

Then again, it would be so nice if even one of them was to surprise me by doing something intelligent.

It is only after we prove the existence of these people, the Giants, that we can focus on getting laws passed to protect them. This can only be done through the United Nations, as the Giants have no knowledge or concept of international borders. Besides, the local authorities have other priorities.

Other Species of Human

In support of my argument, that we are not the only species of human on the earth, may I suggest that another species of human still exists, on the Island of Flores. These people are referred to as Homo Floresiensis, commonly referred to as "Hobbits".

In fact, Dr. Gregory Forth, a highly respected anthropologist, has recently suggested that these people may still exist! He is basing his opinion on those who claim to have seen these Hobbits.

It is thought that the brain size of the Hobbits, on the Island of Flores, is around 400cc, the same size as the brains of chimps! Yet the Hobbits are unquestionably human! It is very likely that Dr. Forth will soon prove that they still exist, so that will remove all doubt!

The fact is that "human" is not determined by the size of the brain! Now if only the scientists could face that fact!

For this, Dr. Forth deserves a great deal of credit. He is challenging a scientific theory, which is widely accepted. Apparently, the scientists are prejudiced against all other species of human!

This prejudice helps to explain the attitude of the scientists, concerning a recently discovered species of human, in South Africa, which lived possibly 200,000 years ago. The scientists have given that species the name of Homo Naledi.

Arguments have been made that the brain size of Homo Naledi, that of 600 cc, in that "cc" stands for cubic centimetres, is too small to be that of a human. This argument ignores the fact that Hobbits were, and likely still are, human! Even with their 400cc brains!

The discovery of Homo Naledi was made quite by accident, as a couple of amateur cave explorers, those who engage in the recreational pastime of exploring wild cave systems, stumbled upon the bones of human beings. These remains were found in a remote cavern at the end of a system of caves, referred to as the Rising Star System, and was almost completely inaccessible. This is to say that the entrance to the caverns are very narrow. Only the most slender individuals are able to access the caverns. A second cavern, equally inaccessible, was later found, also containing the remains of humans.

A scientific expedition to excavate the remains was organized by Dr. Lee Burger, a scientist who specializes in the field of paleo anthropology.

For the sake of those who are not scientific experts, I should mention that paleo refers to people who lived many years ago, and anthropology refers to the study of human beings. Quite reasonably, the scientists who specialize in this field are referred to as paleo anthropologists.

Burger immediately organized a team of young, slender, dedicated scientists to engage in the extremely dangerous task of excavating those bones from the caverns. Not too surprising, all of these scientists were female, and the nearly complete skeletons of eighteen individuals were removed from these chambers. No one was seriously injured during this process.

As there was no indication of predatory damage to the bones, it was clear that predators did not drag the bones into the caverns. Also, the bones of no other animals were found in the caverns, so that ruled out the possibility of the bones being deposited as a result of flooding.

The only remaining possibility is that these remains were placed in those caves deliberately. Those caverns are effectively a burial ground of an ancient species of human.

Of course there is no shortage of scientific souls who dispute that conclusion. They would have us believe that apes dragged the bodies of their deceased deep underground, in complete darkness, through caverns and tunnels which were supremely dangerous, in order to place them in remote caverns. Any misstep could have proven fatal and in the darkness, it was all too easy to get lost.

As well, at least one depression was dug into the floor of the cave, and several members were placed in that hole. Homo Naledi also carved some art work on the walls of the chambers. This behaviour is not characteristic of apes.

It is my opinion that Homo Naledi was human, and I suspect they navigated their way through those underground passages with the aid of a light source, possibly torches. It is possible that the smoke from those torches can still be detected on the roof of those caves. There is no mention in the scientific literatures of anyone checking for that soot.

The bones have been extensively studied, and it has been determined that these apes were definitely bipedal, complete with features which were a "curious mixture of ape and human". This is to say that it walked on two legs but had also evolved physical features to enable it to spend a great deal of time in trees. This is precisely as I would expect, considering that it is a primitive species of human.

In a previous article, I have argued that in order to become human, an ape first has to become bipedal, so that the the front paws are freed up for labor. This is the correct Marxist understanding of the evolutionary process which an ape goes through, in the transition from ape to human.

As previously mentioned, in the process of labor, of crafting ever more elaborate tools, the front paws are gradually transformed into hands, which is to say the opposable thumb is created, so that they are able to touch their fingertips with their thumb. As well, corresponding changes also take place to the body, and the ape gradually becomes human.

In particular, Engels explained this quite clearly in his excellent book, The Origin of the Family, Private Property and the State. The period of transition from ape to human is referred to as the lower stage of savagery, the "childhood of the human race". This period may have lasted many thousands of years, and the main result of this is the development of articulate speech.

During that time, Engels stated that people were "partially at least a tree dweller, for otherwise their survival among huge beasts of prey cannot be explained." That is the reason I am not surprised that they have a mixture of physical characteristics, of both ape and human.

That is a fundamental tenet of Marxism, and as I have gone into this in more detail in a previous article, I will not go beyond this brief summary.

This is to stress the point is that all species of human, which have ever walked the earth, have been, and are now, bipedal apes with opposable thumbs.

Burger reports that a number of complete hands have been excavated from the caverns, and it is his belief that these hands "could have made and manipulated tools". That is very close to saying that these hands had the opposable thumb.

As that is the case, it proves that Homo Naledi was in fact human, so there is no point in arguing that an ape with such a small brain cannot be human. But no doubt there are many scientists who will argue the point.

Of course the brains of our earliest human ancestors were small, the same size as apes, as they were descended from apes. Engels goes on to quote Morgan, an anthropologist for whom Engels had the utmost respect: "When two advancing tribes, with strong mental and physical characters, are brought together and blended into one people by the accidents of barbarous life, the new skull and brain would widen and lengthen to the sum of the capabilities of both."

It is assumed that this species of human lived at the same time as our human ancestors, and then died out. I am not so sure. As yet I am aware of no DNA test to prove this, and the fact is that different species of human have interbred over the centuries. None of us can claim to be a pure bred species. Instead, we are a mixture of Neanderthals, Denosivan and Homo Sapiens.

Most of those bones were recovered from one rather small area, less than one square meter, in one chamber. These human remains are being treated with great respect, properly so, as we are effectively raiding the burial ground of people who lived a very long time ago.

This is not to say that we are grave robbers, as such people merely steal from the dead in order to enrich themselves. That is far different from a proper scientific examination, and in a perfect world, after such an examination, the bones could be returned to their place of burial. That of course is out of the question, as the site has been well marked and grave robbers would no doubt merely steal the bones and sell them on the black market.

With that in mind, I can only suggest securing the Rising Star cave system, in order to prevent grave robbers from climbing into those burial grounds and stealing any remaining bones left in those chambers.

Burger is of the opinion that science should be made more accessible to non scientists. He is also of the opinion that, in all fields of science, extraordinary things are yet to be found. Possibly this latest discovery will inspire people to begin searching.

I most emphatically agree.

PART 4

SWIMMING MONSTERS

Fresh Water Whales and Ichthyosaurs

For many years, I have been focused on investigating various "myths" and "legends", in the belief that such legends are based on fact. These include such legends as that of "Ogopogo" and the "Loch Ness Monster", as well as "Under Water Lights".

It was the legend of "Ogopogo", the "Lake Monster" in Okanagan Lake, as well as "Nessie", another "Lake Monster", in Loch Ness, that gave me the most difficulty.

Incidentally, my method involves gathering as much highly reliable information as I can, and then ruling out the impossible. The only remaining alternative, regardless of how highly unlikely that may be, must be a fact! After all, highly unlikely is not the same as impossible!

The fact of the matter is that, for many years, countless people have reported seeing an animal in Okanagan Lake. As most of those people are solid, respectable, hard working common people, I take their reports quite seriously. Yet their descriptions of that animal varies. Some people swear that it has a "hump" on its back, while others maintain that the animal is "long and slender", snake like, but without a hump. Remarkably enough, both are correct!

The animal which has been described as "long and slender" is a whale! Basilosaurus! Twenty meters long! The animal which has been described as having a "hump" on its back, is an ichthyosaur! That "hump" is in fact a fin!

The fact that all scientists dismiss all eye witness reports, as being "products of an overly active imagination", I consider to be nothing short of offensive.

I could not understand how a fresh water lake could support one population of huge predators, never mind two! Then I got a phone call from my friend, who lives beside Okanagan Lake. He has personally seen one of them, strictly by chance, so he knows that "Ogopogo" is something more than a myth!

He also mentioned that his friends, while scuba diving, had seen something very strange. Tracks! This gave rise to that which I refer to as a "bombshell" moment! That which had been puzzling me for so long, suddenly made sense! These animals have legs!

Lake Monsters As Omnivores

This calls for a little explanation. All scientists are agreed that whales once had legs, and used to walk on land. Then, around fifty million years ago, whales took to the water. As they adapted to a life completely in the water, they no longer used those legs. Over a period of countless generations, those whales lost their legs. This is referred to as a typical example of evolution at work. If an appendage is no longer useful, it is gradually "discarded".

There is some truth to this. That is precisely what happened to most species of whale. Most, but not all! There is one species of whale that did not completely adapt to life in the water. That species of whale is basilosaurus. This whale still has legs! They use those legs!

These fresh water whales are also land based! They spend the winter months, as well as the day light hours, inside caves. After sundown, in the summer months, they come out of the water to graze! These whales are predators, but not carnivores! They eat flesh and vegetation! They are omnivores! It is not the Lake that supports them, but the ecosystem! This ecosystem is composed of the Lake, along with the streams that feed the Lake, as well as the adjacent meadows, swamps and forests! Further, as top predator, the health of the ecosystem depends upon the top predator!

In support of that statement, I have previously documented the fact that, in Yellowstone National Park, all of the wolves were once killed by the Park managers. The top predators were removed from the Park! The result was an ecological disaster! In the absence of the wolves, the Park was being turned into a desert! The Park managers were forced to import wolves, in order to restore the balance of nature.

Now to examine the other huge predator in Okanagan Lake, the ichthyosaur. Not a great deal is known about this animal, as the scientists are agreed that ichthyosaurs were reptiles, which went extinct 95 million years ago. They are also agreed that these animals were warm blooded, and gave birth to live young, head first! Such nonsense!

This makes absolutely no sense! Reptiles are cold blooded animals, which lay eggs! It is only mammals, warm blooded animals, which give birth to live young! For that reason, ichthyosaurs cannot be reptiles! They must be mammals!

Further, as they give birth to live young, head first, they have to come out of the water to give birth! Otherwise, the youngsters will drown as they are being born! The only way they can come out of the water, is with the use of legs! Ichthyosaurs are walking mammals!

This is the reason I say that the ichthyosaur is a mammal, and must have legs. It too is an omnivore, nocturnal, also comes out of the water after sundown to graze.

It is also a fact, that highly reliable eye witnesses have reported that this animal exhibits some most unusual behaviour. It has been known to swim beside people! Such behaviour is characteristic of dolphins! I can only suggest that these animals, ichthyosaurs, are nothing other than a species of dolphin! Of course, that remains to be seen.

Both are mammals. Both are predators. Both are nocturnal. Both have legs. Both are omnivores. Both spend the daylight hours, as well as the winter months, inside caves. In the summer months, both come out of the water, to graze. Stranger than fiction! Yet as I have ruled out all other possibilities, this must be the case!

I suspect that whales and dolphins have learned to coexist, in the same ecosystem. Which is not to say that they are "good buddies", as both are top predators. As that is the case, they very likely kill each other, at every opportunity, as is characteristic of top predators. That too, remains to be seen.

Allow me to stress the fact that even though these animals are nocturnal, they are also predators. They have a keen sense of smell! Even while in those caves, they can smell blood! For that reason, they are occasionally seen in the daylight. I mention this for the benefit of women of childbearing age!

One such lady approached me, as she was attacked by the animal, while water skiing in Okanagan Lake. Not a good idea! She is alive today by the Grace of God! Predatory mammals are less likely to attack people, but it does happen! As opposed to reptiles, which prey upon people at every opportunity!

Locating These Lake Animals

We can compare this to the hippopotamus of Africa. According to the internet, these animals eat meat and vegetation. They come out of the water after sundown, in order to graze on vegetation, mainly grass. That is their main source of nourishment. Apparently they cannot tolerate the heat of the sun. It may burn their hides.

It is reasonable to assume that whales and ichthyosaurs have adopted the same survival mechanism. Two reasons for avoiding the sunlight come to mind. The first is that their hides may be very tender, so that the sun may burn them. The second reason is that possibly birds of prey, raptors, may land on their backs and tear out flesh. If those wounds then became infected, it could prove to be fatal.

The task of locating these animals just became much easier! There is no need of boats, sonar, radar, or submersibles with under water cameras and listening devices! A simple trail camera should be sufficient!

The place where they enter and exit the water must be quite distinctive! A large bare spot of ground, commonly referred to as a "slide". If there is a tree close by, then that is a good place for a trail camera! It is very likely that these animals come out of the water after sundown, but before full darkness. As well, they very likely return to the water after sunset, but before dark.

According to the internet, the Penticton Chamber of Commerce, has recently offered a two million dollar reward, for anyone who can offer proof of the existence of Ogopogo.

No doubt, those who work on Okanagan Lake, have noticed the slides on that Lake, leading to meadows, next to the Lake. Those who can afford to rent a boat can easily place a trail camera, on a tree next to that slide. Those who cannot afford to rent a boat, can fly a drone around the edge of the Lake, beside a meadow, and locate those slides.

I can only hope that this information will inspire people to join in the search for "Ogopogo", the animals which I maintain are whales and ichthyosaurs. If true, then you can take part in a major scientific breakthrough. If not true, then you can quite cheerfully point out my mistake!

It may help to think of this as part of the revolutionary uprising, as that is precisely the case. Working people have to become politically active, and that includes challenging scientific theories. This is a fine place to start. Consider this to be part of training, preparation for the Dictatorship of the Proletariat.

Assuming a picture is taken, of a huge animal grazing, there is always the question of precisely which animal! There is an easy way to tell them apart. The ichthyosaurs have a fin on their backs. This fin is usually described as a "hump". The whales have no such fin.

This little detail may be of vital importance, to the people looking for those animals. At least, it could prove to be most profitable!

Whales and Ambergris

The method I use, to investigate the possible characteristics of such animals, those which are thought to be extinct, is to compare them to their closest living relatives. In the case of basilosaurus, it must be similar to another toothed whale, that of the sperm whale.

The two are quite similar, predators, close to the same size and weight. Yet to my surprise, I found that the "poo" of sperm whales is considered to be supremely valuable! Almost as valuable as gold! Forty thousand dollars per kilo!

This poo is referred to as "ambergris", or "amber grease", or "grey amber", and according to the internet, "among the secret scatological world of whale poo traders", is considered to be the "most valuable poo in the world"! No doubt!

Further according to the internet, this ambergris is "prized" by the perfume industry. It is "used as an ingredient in the most expensive scents". It is also occasionally used as incense, an aphrodisiac, and a medicine.

The scientists are agreed that this, the "worlds strangest natural substance", is produced in the digestive tract of sperm whales! That is about the only thing upon which they agree!

Yet as it is produced by sperm whales, could it possibly be produced by basilosaurus whales? As both whales are predators, it is very likely that their digestive tract is similar. It is entirely possible that this ambergris is strictly characteristic of salt water whales, not fresh water whales. It may have something to do with their diet, which is different from that of fresh water whales. Or not! There is one way to find out!

At the time they graze, they also "fertilize" the meadow. Their droppings should be examined. Bear in mind that ambergris has been described as an "elusive, smelly substance", one which is "frequently mistaken for pebbles and pumice". It may have a scent "similar to sandal wood and tobacco", according to the internet.

As for those who object to handling droppings, may I suggest wearing gloves. Then place the disagreeable substance in a plastic container, and send it to a lab.

As these animals are so huge, they very likely graze on all of the meadows adjacent to the lake, or at least, all meadows which are accessible. Unless, of course, the ichthyosaurs have laid claim to that particular meadow!

There is no harm in making the attempt! If nothing else, proving the existence of even one of these huge animals, will mean a major scientific breakthrough! That alone should serve as sufficient motivation!

Loch Ness Monster

Countless members of the public are fascinated with the animal known as the "Loch Ness Monster". This despite the fact that the scientists dismiss all sightings, as "products of an overly active imagination", at best. At worst, they refer to these sightings as "deliberate deception".

Sadly, this deliberate deception does happen. A famous video of that which was billed as "swimming Nessie", is one of those deliberate deceptions. A bad joke!

That in no way changes the fact that countless working people have seen the animal in Loch Ness, that which is commonly referred to as "Nessie". As previously mentioned, it is rarely seen, because the animal is nocturnal. It spends all of the day light hours, as well as the winter months, inside caves. Then again, as it is a predator, it has a keen sense of smell. Even inside caves, it is able to smell blood, and then attacks. That explains the occasional daytime sightings.

Yet the scientists dismiss all such sightings. The scientists could not possibly be more mistaken! They have chosen to disregard solid evidence, which supports the eye witness accounts.

In particular, reports of a "monster" inhabiting Loch Ness, "date back to ancient times". There is even a famous stone carving, by the Pict, depicting a "mysterious creature with fins". As it was carved by people many hundreds of years ago, it is strong evidence that the animal existed, at that time. Further, as those same people went to the considerable effort to carve the symbol in stone, they must have held it in great regard. There is even a more recent account dating from the year 565 CE, concerning St. Columba. The scientists dismiss all this as "Scottish folklore".

More recently, in 1933, a couple spotted a "dragon or pre historic monster", which "disappeared into the water", as was reported in a local newspaper. As these reports seemed to be credible, a big game hunter investigated and found "large footprints", along the shore of the lake. But then zoologists at the Natural History Museum dismissed these tracks as a hoax!

To think that key evidence was dismissed out of hand! Those tracks were almost certainly the tracks left by this animal, as it returned to the water! This animal, the "Loch Ness Monster", has legs! It uses those legs! After sundown, it comes out of the water to graze! It is an omnivore! It eats vegetation, as well as meat!

As legend mentions an animal with fins, it is most likely an ichthyosaur, as those animals have fins on their back. Which is not to say that it could not be a fresh water whale, basilosaurus, as it is entirely possible that both species of swimming mammals exist in Loch Ness, as they do in Okanagan Lake. That remains to be seen.

As previously mentioned, all efforts to locate these animals have failed, both here and in Britain. Yet very soon, we will locate them. There is no need for boats, submersibles, under water cameras or sonar. A simple, motion activated trail camera is all we need. It is a simple matter of attaching the trail camera to a tree, on one of the meadows, close to a slide. As the animals graze, eating vegetation in the moonlight, no doubt the camera will take a picture.

The people who live around those lakes, frequently hear the sounds the animals make, as they enter and exit the lake, as well as at the time they graze. By all means, work with those people. In this manner, common people can take part in a scientific breakthrough.

We do not need a degree in science, in order to work in science! As the scientists are determined to not perform their duty, then it is up to common people to perform that duty for them!

As for those who suggest that it is highly unlikely that the same animals could exist, on two separate continents, so far apart, separated by oceans, may I suggest that there was a time when all of the continents were grouped together in one huge world landmass, referred to as pangea. The descendants of the animals which lived on pangea, are still with us.

Migration As A Means of Limiting Inbreeding

People who live around the huge lakes in North America, can testify to the fact that "something" is in those lakes! I am sure that "something" is whales and ichthyosaurs. No doubt there are populations in those lakes. (I use the word "population", because I have no idea what a group of ichthyosaurs are called!)

This creates a little problem, in the form of incest. Not healthy! These animals have overcome that little problem by chasing the young males away, as soon as they come of age! Probably at the age of four, they are forced to "vacate the premises", to leave the lake, to travel, by river, to other huge lakes, where they can find mates! There is no other way!

But as previously mentioned, the scientists refuse to even consider the possibility of the existence of these animal. Yet they are part of our heritage! We are entitled to our wildlife! We are being robbed of our heritage!

We are also entitled to a clean environment! It is imperative that we, working people, prove that they exist! That is our moral obligation! Then it is a matter of forcing through laws to protect them! I say "force through", because that is the last thing the capitalists want!

These animals can best be protected by cleaning up our rivers and lakes, for a start. The capitalists take great delight in polluting both, by dumping garbage and toxins from their factories, into that water. That is so much cheaper and easier than disposing of it properly!

Yet as a result of the revolutionary movement, certain scientists in Britain have stepped up, and are challenging the generally accepted scientific belief, that there is no "Loch Ness Monster"! In so doing so, they are risking career suicide! They are placing their careers "on the line"! It is entirely possible that they may be fired, from any job they have, working in any field of science! For this, they deserve full credit!

That being said, it must also be pointed out that the act of doing that which has been done countless times before, while expecting a different outcome, is the very definition of insane! Yet that is precisely what they are doing!

According to the internet, in August of this year, the Loch Ness Centre and a volunteer research team called Loch Ness Exploration, organized a search for "Nessie". Hundreds of volunteers showed up for a weekend quest. The equipment included drones fitted with infrared cameras, and a hydrophone. Most encouraging!

The fact that so many people showed up, volunteering their time, is an indication of the strength of the revolutionary motion. Working people are becoming active! Gaining experience! Challenging both the government officials and the scientists! Determined to prove that which they know to be a fact! The existence of "Nessie"!

Of course, the search failed to turn up any evidence of the existence of Nessie, because all were focused on looking in the wrong place, at the wrong time! They were all looking in the water, during the daytime, when the animal- or animals! are inside caves! After sundown, they all "went on strike"!

The volunteers, all common people, no doubt, cannot be faulted for this! On the contrary, they deserve full credit for becoming active, taking part in challenging a scientific theory which is thought to be "sacred"! Not to be challenged!

It is the scientists, those who organized this search, who should have at least considered the possibility that these animals are nocturnal.

As for the suggestion of "chumming", throwing bloody meat into the water, I strongly advise against that! That could lead to an environmental disaster! It would very likely lead to a "feeding frenzy", in which the animals, possibly whales and ichthyosaurs, would attack each other!

The reason I use the expression "environmental disaster", is because these two animals are "top predators"! The health of the ecosystem depends largely upon the top predators! The experience of Yellowstone National Park proves that!

The point is that we must first prove the existence of whales and ichthyosaurs, the top predator in these huge lakes, and then protect them. The health of the whole ecosystem depends on this.

It is somewhat unfortunate, that during the whole month of August, I was focused on a different project, one which demanded my undivided attention. For that reason, I was not aware of the plan to look for "Nessie" during a weekend in August. I stumbled upon an article, concerning that event, after it was over. However, as I consider it to be so important, I will reproduce it here:

"New Generation of Loch Ness Monster Hunters In Quest to Finally Solve the Mystery!

"Armed with drones, infrared cameras and a hydrophone, dozens of volunteers are converging on a Scottish Lake to search for a legend no one else has found before.

"Biggest search in fifty years!

"Quest weekend- August 26, 27 has drawn interest from around the world and is being billed as a chance for a new generation of monster hunters to help uncover the truth

"Thermal drones equipped with infrared cameras at night to scan the surface of the Lake in the Scottish Highlands and to submerge a hydrophone to listen for any 'Nessie' -like calls under the water. There are more than one hundred volunteers.

"Group Loch Ness Exploration"

I repeat, it is most encouraging to see both professional people, as well as common people, taking part in challenging a "sacred scientific theory"! It is very likely that all involved, devoted their time, for free, in the interest of science.

Even though the weekend search was over, I decided to post a comment on that web site:

"As the author of Bird From Hell and Other Mega Fauna, Third Edition, I can testify to the fact that the Loch Ness Monster is almost certainly the same animal which is located in Okanagan Lake, commonly referred to as Ogopogo. In fact, there are two huge animals, very similar. Both are mammals, nocturnal, both have legs, both are predators but not carnivores. Both spend the daylight hours and the winter months inside caves, except when they smell blood. Both come out of the caves, and out of the water, in the summer months to graze. This makes sense as a fresh water lake cannot possibly support a population of huge animals. There is simply not enough nourishment in the lake to support such a population of huge animals. May I suggest flying a drone around the edge of a meadow, adjacent to the lake, looking for a slide, which is a bare patch of ground where the animal enters and exits the lake. If there is a tree close by, then attach a wilderness camera to the tree. It is very likely that the animal comes out of the water after sundown, but before full darkness. Even if not, then perhaps in the moon light, you will capture a picture of the animal. Either a whale or an ichthyosaur. Good Luck"!

I am confident that people around Loch Ness, as well as those around Okanagan Lake, both professional and amateurs, will soon take my advice and locate those huge mammals. I have complete confidence in you!

Sea Monsters

Let me start by saying that, as previously mentioned, the scientists are convinced that no less than five orders of reptile have gone extinct, even though they also maintain that *there are no mass extinction of reptiles!* In addition to the flying reptiles, pterosaurs, it is their opinion that there was a mass extinction of four orders of swimming reptiles.

These include the long necked plesiosaur, the short necked plesiosaur, the mosasaur and the ichthyosaur. Swimming reptiles, one and all. Or so they say! Yet such is not the case!

In previous articles, I made the mistake of taking the scientists "at their word", in that they claim these animals are reptiles. In fact, just as ichthyosaurs are mammals, so too, mosasaurs and plesiosaurs are also mammals. They give birth to live young. For that reason, the techniques I suggested, in previous articles, which involved looking for the eggs they laid, rather than looking for the animals, is faulty. In the interests of correcting that mistake, I wrote a separate article, in a different book. I have yet to select a title for that book.

Perhaps the simplest way to correct that mistake, is by reproducing that article, in its entirety, in this book also. It is titled:

In Search of Mosasaurs and Plesiosaurs

The title of this chapter may seem to be rather strange, the subject completely out of place, in a book such as this, as this quest has nothing to do with the class struggle. Besides, "everybody knows" that those huge, pre historic reptiles went extinct, many millions of years ago.

Such an attitude is completely understandable, but mistaken, and calls for an explanation.

The class struggle between the working class, the proletariat, and the monopoly capitalist class of multi billionaires, the bourgeoisie, is not limited to the political arena. On the contrary, it extends to all aspects of life. This includes the cultural, academic, historic and scientific fields. In fact, as the various fields are so closely tied together, with the class struggle, in so many different ways, it is sometimes difficult to separate them.

The point being that it is necessary to fight the capitalists, on all "fields of battle", including that of science. In this article, we are focused on the scientific field of palaeontology, which is defined as "the study of life forms that existed in previous geologic periods, as represented by their fossils".

Just because certain animals existed many millions of years ago, does not mean they do not exist now! Absence of proof, is *not proof* of absence! All too often, it just means that we have not been looking! Or at least, not looking in the right places!

In previous articles, I have laid out the procedures to be followed for locating various land dwelling species, which are classified as extinct, even though they are not. I strongly suspect that these "salt water reptiles", mosasaurs and both species of plesiosaur, long necked and short necked, are also very much alive. As they are located in the ocean, the task of finding them is far more difficult.

We can begin our investigation with that of the mosasaur. According to the scientists, it was 17 meters long, and weighed ten tons. As they phrase it,

"Mosasaurs were top predators of the world's oceans, and would eat anything they could catch".

Yet those same scientists state that these animals went extinct approximately sixty five million years ago, at the same time the dinosaurs went extinct. A remarkable coincidence, and in my opinion, completely bogus. More on that subject, later on in this article.

As for plesiosaurs, it is the scientific opinion that they were of similar size, and also went extinct, at the same time the mosasaurs went extinct. Or possibly a few million years earlier, as different scientists say different things. Yet all are agreed that these huge, pre historic, "salt water reptiles", are indeed extinct.

Another one of their beliefs, is that there was a mass extinction of dinosaurs, sixty five million years ago. They just cannot seem to agree upon a proper scientific definition of a dinosaur, or of the cataclysmic event which caused their extinction. Different scientists say different things. No wonder the public is confused!

Incidentally, in previous writings, I have made my position quite clear. Briefly stated, all of the animals which are classified as dinosaurs, laid eggs. Those which had feathers, were birds. Those without feathers, were reptiles. There was no mass extinction of either birds or reptiles! The theory of the mass extinction of dinosaurs is nonsense!

For the benefit of those who are skeptical, allow me to point out that perhaps the most famous "dinosaur" of all, was Tyrannosaurus Rex, or T Rex. All scientists are agreed that T Rex went extinct at the same time that all dinosaurs went extinct, around sixty five million years ago, possibly as a result of a "killer asteroid from outer space".

Yet those same scientists are also convinced that the common barnyard chicken, is a descendant of T Rex. Just how a species can be completely wiped out, practically overnight, and yet still give rise to descendants, is a paradox they do not even attempt to explain.

This is to say that the scientists, working in the field of palaeontology, have managed to create a complete muddle. This is not too surprising, as it is characteristic of the bourgeois, according to Lenin.

It is up to working people, to straighten out this confusion. After all, we are entitled to our heritage, and wildlife is part of that heritage.

Now to return to the topic of the supposed extinction of "salt water reptiles". As I believe in giving credit where credit is due, we can start with Dr. Robert Bakker. He is highly respected, widely considered to be the foremost authority, in the field of palaeontology. His book, The Dinosaur Heresies, is widely considered to be the "bible" of palaeontology.

In that book, he had a few words to say, concerning the supposed "mass extinction" of "salt water reptiles": "As dinosaurs were snuffed out at the end of the Cretaceous, the great sea lizards, and the snake necked plesiosaurs were also dying out, as were a host of large and small invertebrates, from coral like oysters to shelled squid and microscopic plankton".

Bakker would have us believe that, at the same time the dinosaurs were being "snuffed out", no doubt on land, the "great sea lizards", likely a reference to the mosasaurs, as well as the "snake necked plesiosaurs", "were also dying out". Just how this supposed "land extinction", spread to the ocean, so that the top predators, the "great sea lizards", also went extinct, is not explained. They certainly did not starve to death!

Apparently, we are supposed to take this theory, that of the mass extinction of "salt water reptiles", as an "article of faith". Not this child!

In my opinion, all personal beliefs, articles of faith, are to be respected. By contrast, all scientific beliefs, which are nothing other than theories, are to be challenged.

Those who support these scientific theories, which are being challenged, should not take this personally. On the contrary, they should embrace such challenges. After all, we are all human, and we all make mistakes. We should be grateful to anyone who can point out our mistakes.

That is not the attitude of the vast majority of scientists. On the contrary, they have been corrupted with the bourgeois influence. They consider any such criticism as a challenge to their authority. Any University student, in any field of science, who questions any scientific theory, presented in any course of science, is certain to fail that particular course, upon which he or she is enrolled. As well, any scientist, working in any field of science, who questions any "generally accepted" scientific theory, is sure to commit "career suicide".

That is the current state of science, and as such, is completely unacceptable. It is up to the working class, the proletariat, the "only consistently revolutionary class", according to Lenin, to straighten out this mess.

With that in mind, for the benefit of those workers who have little or no scientific background, we can start with a definition of mammals and reptiles.

The simplest possible definition of a mammal, is that of "any animal in which the female gives birth to babies, not eggs, and feeds them on milk, from her own body".

That definition gives us a place to start, and is largely true. The exception to that rule can be found in Australia, in which two species, that of platypus and echidna, can be found. They are classified as mammals, as they are warm blooded animals, and the females feed their babies with milk, from their own bodies. Yet those same females lay eggs!

These animals are known as "monotremes", an "ancient order of mammals", which have characteristics of birds and reptiles, as they lay eggs.

This brings us to reptiles, which are defined as "cold blooded animals, lay eggs, and has a body covered with scales or hard parts". As well, cold blooded animals cannot generate their own body heat, in contrast to mammals and birds, which can do just that. Most reptiles rely mainly upon the heat from the sun, in order to "warm up".

All scientists are agreed that mosasaurs were reptiles, which is rather strange, as those same scientists are also agreed that they gave birth to live young! They know this, for a fact, because recently, a "mosasaur skeleton containing five unborn young in its abdomen was discovered in South Dakota", according to the internet. (By "skeleton" they are no doubt referring to the fossilized remains.) So how can these animals, which give birth to live young, be classified as reptiles?

Of course, it is entirely possible that mosasaurs are indeed reptiles, which have characteristics of mammals. And in fact, the scientists are careful to state, quite emphatically, that they are reptiles, as they have "certain characteristics of reptiles". Just what these "certain characteristics" are, they are careful not to say.

On the other hand, I am certainly not convinced. The existence of monotremes proves, beyond any shadow of a doubt, that certain species of mammals have characteristics of birds and reptiles. It stands to reason that certain species of reptiles, can also have characteristics of mammals. Yet just as monotremes are supremely rare, so too, the possible existence of reptiles with characteristics of mammals, must also be exceptionally rare.

This brings us to the subject of plesiosaurs, also referred to as "marine reptiles". Here too, the latest evidence, that of fossilized remains, has the scientists convinced that they too, reproduced by giving birth to live young. Or at least, "some plesiosaurs gave birth to live young", as they so carefully phrase it. They also suspect that those same animals "mothered" their youngsters, in much the same manner of dolphins and whales.

I suspect they are absolutely correct, in the sense that those animals "mother their young", but only because the plesiosaurs are mammals, just as whales and dolphins are mammals. For that reason, they have characteristics which are common to mammals. Naturally, that remains to be seen.

It is further significant that all scientists are agreed, that both mosasaurs and plesiosaurs lived in low lying, coastal areas. The reason I say this, is because it is entirely possible that they still live in "low lying, coastal areas". That too, remains to be seen!

With that in mind, numerous reports of "under water lights", should be investigated. These have mainly been seen off the coast of California, as well as the Solomon Islands. According to the scientists, these are "most likely bioluminescent plankton".

Very likely. But until we check, we will not know for sure. Experience has taught me not to take anything for granted. Check and double check! Or end up wishing you had!

The reason I say this, is because certain details puzzle me. I am convinced that these animals exist, yet the fact is that sailors rarely report spotting them. Or at least, not to my knowledge. As they are huge, they are difficult to miss.

Unless of course, they are nocturnal. It is entirely possible that they spend the day light hours inside caves. This may lead to difficulties during mating season, in which the males have to display, in order to impress the females. The difficulty being that the females cannot see the males display. Unless of course, the males have evolved the ability to glow. That is the way pterosaurs have resolved the problem of being nocturnal.

This idea is not as "far fetched", as it may first appear. As I have previously documented, the fresh water whale, basilosaurus, as well as the ichthyosaurs, which are classified as reptiles, even though they are mammals, are both alive and well. They are also nocturnal, as well as omnivores, so that they come out of caves, after dark, and go onto meadows next to fresh water lakes, to graze. Is it possible that these salt water animals behave in a similar manner? Could they be spending the daylight hours inside caves, yet grazing in the meadows next to the ocean, after sundown?

Just as I maintain that ichthyosaurs are mammals, not reptiles, may I also suggest that both mosasaurs and plesiosaurs are also mammals. Not that I am convinced of this, which is the reason that it is merely a suggestion.

The reason I say this, is because reptiles do not hesitate to prey upon humans. As far as they are concerned, we are merely prey animals. Dinner! Yet there are no reports of people being preyed upon by such animals.

By contrast, predatory land dwelling mammals, such as bears and wolves, are less likely to attack humans. It certainly happens, but far less frequently. As mammals, they are far more intelligent than reptiles, so that their instincts are to avoid killing us.

This brings me to the subject of predatory swimming mammals, and in particular of orcas, commonly referred to as "killer whales". In fact, they are a huge species of dolphin, and superb predators. They even prey upon hammer head sharks! Yet they have never been known to attack humans!

Here too, as intelligent mammals, it is very likely that their instinct is to avoid killing humans. This is a good thing!

Assuming that mosasaurs and plesiosaurs are mammals, it explains the reason they are not preying upon people. Their instincts prevent this. Further assuming them to be nocturnal, it stands to reason that they may well have legs. How else can they go back and forth into caves? It further stands to reason that they use those legs. They too, may come out of the water, after sunset, onto the meadows, in order to graze. Omnivores!

If that is the case, then the place where they enter and exit the meadow, is quite distinctive. It is a bare patch of ground, referred to as a "slide". Those who work on the ocean, close to shore, can readily point out these slides. Bear in mind that these animals come out of the water after sundown, but before dark, and return to the water before sunrise, but in the daylight. A fine place for an inexpensive trail camera! Well within the price range of most workers!

Incidentally, in previous articles, I made the mistake of "taking the scientists at their word". As they consistently refer to these huge, salt water swimming animals, as reptiles, I believed them. For that reason, I failed to conduct a proper investigation. This resulted in my suggestions for locating these animals, based upon the eggs that reptiles lay.

As we live in a class society, under a state of monopoly capitalism, there are two ideologies. One is bourgeois, and the other is proletarian.

Lenin stressed this fact in his work, What Is To Be Done? As he phrased it, "Since there can be no talk of an independent ideology being developed by the masses of workers in the process of their movement, *the only choice is*: either socialist or bourgeois ideology. There is no middle course (for humanity has not developed a 'third ideology', and moreover, in a society torn by class antagonisms, there can never be a non class or above class ideology)". (italics by Lenin)

It is significant that Lenin refers to the proletarian ideology as being "socialist", in stark contrast to the bourgeois ideology. It is further significant that the bourgeois ideology has extended to all fields of science.

Even though I made a mistake, it was relatively minor, and as we are all human, we all make mistakes. According to Lenin, the important thing is: "Frankly acknowledging a mistake, ascertaining the reasons for it, analyzing the conditions that have led up to it, and thrashing out the means of its rectification".

This article is to be considered as a correction to that mistake.

Now it is up to working class people to assist, in the search for these magnificent animals. The scientists are certainly not about to "lift a finger", as they are so completely consumed with the bourgeois ideology.

As previously stated, we have to fight the capitalists, on all fields. That includes the various fields of science.

PART 5

SCIENTISTS AND SCIENTIFIC THEORIES

Challenging Scientific Theories

This brings me to the subject of scientists. They are certainly not "common people", "members of the public". On the contrary, they are highly trained professionals. Yet the behaviour, of all too many of them, are not professional.

Perhaps it would be best to start by considering the actions of those who study the fossilized remains of animals, which lived possibly many millions of years ago. Such scientists are referred to as palaeontologists.

Several of them have written books, as their research has led them to certain conclusions. These scientific conclusions, or beliefs, are referred to as theories. All too many of these theories are presented as *facts!* These *theories* are not facts, but *beliefs! Scientific beliefs!* As such, these theories are meant to be *challenged!* This is the only proper application of the scientific method! Such a challenge should not be taken personally! Yet all too often, that is the case! Those scientists who are offended by any challenge to their theories, are acting in a manner which is non professional.

Allow me to stress the fact that the only way in which science advances, is through the method of challenging scientific theories!

As a means of stressing the importance of challenging scientific theories, consider the fact that until quite recently, it was thought that we lived at the centre of the universe. Of course, we now know differently, but only because scientific pioneers challenged the accepted theories. They "blazed the trail", so to speak, and we would do well to honour their memory, by following in their footsteps!

This is not happening! In fact, there are certain books of science, which contain *certain scientific theories,* which are *not to be challenged!* I refer to these as "sacred scientific theories". Then again, I have my own name for them. Scientific fairy tales!

As can be well imagined, this does not endear me to the scientific community!

As a consequence of these "sacred scientific theories", *any* student of science who questions *any* of those theories, presented in *any* science course, is certain to *fail* that course! Such a student is *not* allowed to earn a degree in science!

It is also a fact that any scientist who *challenges* those theories, commits "career suicide"! Such people are not allowed to *earn a living* in their chosen field of science! Any work they perform in science, must be *at their own expense!*

For that reason, both students and scientists, have learned to *memorize* scientific theories, rather than *challenge* them! It is the price to be paid for being *successful!*

This can only be referred to as a "retrograde trend" in science! We are going backwards!

Our current scientific situation is beginning to resemble the state of science, at the time which is commonly referred to as the Middle Ages. At that time, the people who were in power, were convinced that we lived at the centre of the universe. They also thought that the bible was all that was needed! God help anyone who disagreed! Such unfortunates were executed!

Proper Scientific Method

It was under those conditions, that Kepler, Copernicus and Galileo risked their lives, in the pursuit of science! They were the first to challenge the theory that the earth was at the centre of the universe! They made a detailed study of the night skies, over a period of many years, tracking the location of all visible objects! To be caught, was to be executed!

At a later date, it was Newton who gathered the work of those scientific pioneers, and brought it all together, in his three laws of motion. As Newton stated, he had merely "stood upon the shoulders of giants". This gave birth to the modern scientific method!

Since the time of Newton, the state of science has regressed. Once again, scientific theories are meant to be memorized, not challenged!

Yet the revolutionary movement, in various highly industrialized countries of the world, has spread to the scientific community. This is an indication of the strength of that movement! Several scientists are risking their careers, in the pursuit of science!

In particular, at least one scientist in Britain, and possibly several, dared to organize an expedition in search of the Loch Ness Monster, referred to as Nessie. This is most significant, because the "non existence" of such an animal, is a sacred scientific theory, which is *not* to be challenged!

The scientist, or group of scientists, who organized that search, deserve full credit! They took a stand, on principle! They are preforming their duty! At the risk of losing their careers!

The fact that the search turned up nothing, is secondary. They were looking in the wrong place, at the wrong time! Now that they know where to look, no doubt next time, they will be successful.

Another scientist who has been touched by the revolutionary movement, is the fellow who is looking for a separate species of human, Homo Floresiensis, otherwise known as "Hobbit". That is Dr. Gregory Forth, an anthropologist.

He too, is challenging another "sacred" scientific theory, which is that we are the one and only species of human, currently alive today. Such is not the case, and no doubt Dr. Forth will soon prove this.

There are certain scientific theories which can be proven to be false, quite easily, possibly without the input of any scientist. On the other hand, any such input, from any scientist, is most welcome.

Sacred Scientific Theories

Earlier in this book, I have documented the fact that numerous species of huge animals, which were thought to be extinct, still exist. As well, I let people know precisely how to locate them. Proving the existence of those animals will disprove several "sacred" scientific theories. This is well within the abilities of working people, members of the public, by whom I mean those who are not professionally trained scientists. Then again, we can only hope that certain scientists assist them. They can certainly learn from each other! The scientists can learn from the practical knowledge and skills of the workers, and the more advanced workers can learn to pursue a career in science, if they so choose.

Yet there are other "sacred" scientific theories, that require the expertise of highly trained scientists, in order to disprove them. Of course, the more advanced workers can be most helpful.

Perhaps the most cherished of all the "sacred" scientific theories, is that of the "mass extinction of dinosaurs". This bit of nonsense was first put forward by Luis Alvarez, a famous scientist, a physicist, who suggested that a huge meteor hit the planet, perhaps sixty five million years ago. He further suggested that, as a result of this impact, all of the dinosaurs had simply dropped dead.

Alvarez was a fine scientist, highly respected. The work that he did in his chosen field, that of physics, was excellent. For that reason, people listened closely to that which he said.

That was somewhat unfortunate, as he was also human. In fact, all too often professional people make the mistake of expressing their opinion, concerning a subject of which they know nothing. Alvarez knew nothing about dinosaurs. That meteor, of sixty five million years ago, certainly did *not* wipe out a great many species of animals, commonly referred to as "dinosaurs".

Alvarez was very likely correct, when he suggested that a huge meteor struck the planet, many millions of years ago. Such meteors are rocks from outer space. They are constantly striking the planet. Most of them are the size of a grain of sand, and immediately burn up in the atmosphere. They are commonly referred to as "shooting stars".

Others are bigger, and hit the earth, occasionally doing considerable damage. Then there is the odd huge rock, which may cause wide spread devastation. That is a far cry from causing the mass extinction of countless species of animals!

Yet this theory was embraced by countless members of the public. Who can blame them? It has a certain romantic appeal. The idea is that all huge, terrifying, prehistoric animals, were killed by a rock from outer space! In this manner, the world was made safe for people! If nothing else, the theory sold a great many magazines!

The implication, that the mass extinction of dinosaurs was an Act of God, is quite agreeable to most common people. It is quite preferable to the belief in the theory of evolution!

That in no way changes the fact that the theory of the mass extinction of dinosaurs, is merely a fairy tale! Yet what are dinosaurs?

For many years, scientists have been aware that dinosaurs laid eggs. The discovery of the fossilized remains of countless dinosaurs eggs leaves no room for any doubt! Yet the laying of eggs is characteristic of reptiles, as well as birds. But it is only birds that have feathers. The trouble being, that feathers do not fossilize. So how to prove that dinosaurs had feathers, and were therefore birds?

Fortunately, it is not just the fossilized eggs of dinosaurs which have been discovered. The fossilized remains of the hides of various species of dinosaur have also been found. This is significant, because the "feather follicles" -to use the scientific term- from which all feathers grow, have also fossilized.

With feather follicles in mind, recently the scientists had the brilliant idea of examining the fossilized remains of hides, of various dinosaurs, under a microscope. To their astonishment, they determined that these fossilized hides did indeed have fossilized feather follicles. Further, from the size, shape and distribution of these feather follicles, they are able to determine the size, shape and even the colour of the feathers, of these dinosaurs. Not only does this prove that dinosaurs were birds, but we also know the colour of their feathers! As birds certainly still exist, and birds are dinosaurs, then dinosaurs cannot be extinct! This proves that the theory of the mass extinction of dinosaurs is a myth! A fairy tale!

This in no way changes the fact that, as Dr. Robert Bakker stated in his book, Dinosaur Heresies, "dinosaurs are incontrovertibly dead". This despite the fact that he also states that "birds are dinosaurs"!

Dinosaurs cannot be both dead and alive! Yet Bakker is recognized as an authority in his field. His book is considered to be the "bible" of palaeontology!

For that reason, modern day scientists have been forced to resort to verbal gymnastics, in an effort to reconcile the irreconcilable, to explain that even though dinosaurs are extinct, birds are dinosaurs! The alternative is career suicide!

Along with the theory of the mass extinction of dinosaurs, is the theory of the mass extinction of flying reptiles, technically called pterosaurs, commonly referred to as pterodactyls. I use the expression "along with", because some scientists have classified these flying reptiles as dinosaurs, while others do not. Yet all are agreed that they went extinct at the same time that the dinosaurs went extinct.

These flying reptiles are certainly not extinct! Which is not too surprising, as all scientists are agreed that there have never been any mass extinctions of reptiles! This despite the fact that they also maintain that there have been no less than five mass extinctions of reptiles!

They cannot have it both ways! Feel free to face the fact that flying reptiles are not extinct! Further, they are located on six continents of the world, which means that they are not in the Antarctic. It remains to be seen, if they have managed to spread to various islands. As I have document this in a previous article, there is no need to repeat it here.

Global Warming

The next item on our "hit parade", is that of "climate change", in the form of "global warming". All scientists are agreed that the industrial revolution gave rise to the whole sale burning of "fossil fuels", and that this, in turn, is causing the world to become considerably warmer. This neat little theory conveniently overlooks a few little details.

The fact of the matter is that the climate is constantly changing! This has been going on since the creation of the planet, and if nothing else, is responsible for the fact that we have four seasons! We have no reason to expect the climate to remain constant! That is not about to happen!

This is not to say that the climate changes equally, on all the continents, at the same time. It does not. Different continents experience different rates of climate change.

As mentioned earlier, the continent of North American recently experienced no less than three ice ages. On three occasions, the continent cooled off, to the point that glaciers covered the continent. Also on three occasions, the continent warmed up, so that those glaciers melted.

Continental climate change! It just so happens that we are fortunate to be living at a time in which North America is warm.

Walking On the Moon

This brings us to the most passionately defended bit of scientific hog wash, which is the claim, put forward by various scientists, as well as all government officials, that men *walked on the moon! Nonsense!*

It is Adolf Hitler who is credited with making the statement that, "If a lie is big enough, and repeated often enough, then people will believe it." The lie that "men walked on the moon", certainly qualifies as one of the all time great lies! A true whopper!

But now let us consider the facts which make it *impossible* for people to travel to the moon.

First of all, the earth is a giant magnet, and as such, creates a huge magnetic field, which circles the planet. This is a good thing, to put it mildly. Without that magnetic field, we would not be here!

In scientific jargon, this magnetic field gives rise to something referred to as the "Van Allen Belt", which wraps around the earth, much as a protective blanket. This Belt, or blanket, blocks most of the radiation from the sun, which allows life to exist on our planet. Without that Belt, we would be bombarded with massive doses of radiation.

Yet the scientists would have us believe that astronauts passed through this Van Allen Belt, on their trip to the moon. If that was the case, they would have been exposed to massive doses of radiation, which would have quickly killed them. Perhaps not immediately, but within a few days.

No doubt, many skeptical readers will check on the internet- as I did- in the interest of conducting a proper investigation. With that in mind, may I suggest clicking on the article titled: "Yes, Apollo Flew Through the Van Allen Belts Going to the Moon."

I recommend this article because it provides a very clear cut explanation of the reason that it is not possible to fly men to the moon! Feel free to disregard the title, as the facts presented within the article proves precisely the opposite. I repeat, it is simply not possible, to fly people to the moon!

In this video, which runs for a length of eleven minutes, seven seconds, a lovely young lady, a fine actor, explains that "the earth, which is nothing other than a huge magnet, creates a magnetic field. This magnetic field, which is referred to as the Van Allen Belt, acts as a protective envelope", one which "traps most of the radiation from the sun".

An instructive drawing is thoughtfully provided, as a means of illustration. It is further explained that without this "protective envelope", the "earth would be bombarded with massive doses of radiation, which would kill all life on earth". Or at least it would kill us!

Excellent! So far so good!

She went on to say that the "scientists of the time", fifty years ago, "were well aware that the only way the astronauts could get to the moon, was by passing through this Van Allen Belt, and in the process, soaking up massive doses of radiation". Of course, after they passed through the Van Allen Belt, "they would then be exposed to even more radiation, all of which was sure to kill them". This too is true!

Once again: Excellent!

That explanation, which is stated in very clear, simple terms, should provide non technical people, and most working people are non technical, with a basic understanding of the importance of the magnetic field of the planet. It also explains that it is impossible to fly people to the moon, without getting killed!

Bear in mind that this young lady is not a physicist, but an actor. No doubt she was paid to read a script, probably with the help of a teleprompter, and she did a fine job! If I was working in the field of entertainment, I would not hesitate to hire her! She knows how to deliver a speech! She certainly earned her pay!

The point being that we cannot hold her responsible for the nonsense which followed! She was merely reading from a teleprompter!

It is very likely that the scientists were forced to hire this young lady, as they could not find any scientist who was prepared to stoop to the hypocrisy of reciting the nonsense which followed! At least, not with a straight face!

It was the scientific writers who deliberately led with a few basic facts, followed by a vast amount of detailed technical information, in an attempt to confuse people. Buried within that mountain of gobbledegook, was a jewel of a detail, which is referred to as "shielding of the space craft".

I can only stress that she was reading from a teleprompter. She went on to explain that, "in order to protect the astronauts from the radiation which was sure to kill them", there was a "Radiation shielding to the space craft, especially to the main command module, where the crew would be spending most of their time".

Nonsense!

The scientists would have us believe that the astronauts went to the moon, through the Van Allen Belt, and were protected from the radiation of the sun, by a "protective shield". *Such a protective shield does not exist!* There is no such substance! It does not exist now, and did not exist fifty years ago! There is no way that astronauts could possibly have gone to the moon, without being exposed to massive doses of radiation! That radiation would have killed them!

Yet to this day, the scientists, and all government officials, insist that "men walked on the moon"! This despite the fact that it cannot be done, without a "protective shield", which does not exist! Duh!

It is quite possible that this lie may go down in history as the ultimate absurdity!

Retrograde Trend In Science

Sadly, the state of science has regressed to a point which can be compared only to that which existed before Newton and Darwin. This is to say that modern scientific enquiry no longer exists!

The problem is one of capitalism, at least capitalism in its final, most rotten stage of monopoly, technically referred to as imperialism.

As Lenin stated, imperialism is "reaction, right down the line". The imperialists, in the form of the monopoly capitalists, the billionaires, the bourgeoisie, can and must be overthrown. Until they are overthrown, there will be no progress in any field of science. Or at least, not if the billionaires have their way!

It is the working class, the proletariat, which is destined to overthrow that most reactionary class, of billionaires, and crush them, under the Dictatorship of the Proletariat.

Yet the working class is not aware of itself, as a class, with its own class interests. The conditions of life, of the working class, does not lead to that awareness. That awareness must be brought to the working class, from an outside source.

In particular, they must be made aware of the fact that the existing state apparatus must be smashed, at the time of the Revolution, and replaced with a new, Proletarian state apparatus, in order to crush the capitalists, as they make every effort, after the Revolution, to restore their "paradise lost". This state apparatus is referred to as the Dictatorship of the Proletariat. We will know we are being successful, when that expression becomes common place.

Now that the working class is cultured, it is quite capable of reading and understanding that most important work of Lenin, State and Revolution. That lets workers know, precisely what has to be done, at the time of the revolution. Yet the input of middle class intellectuals, would be most welcome.

With that in mind, I have a word of advice for a group of people, those who are middle class intellectuals, self proclaimed Socialists. I understand that most of you think that Socialism is a good idea, but simply not possible. Your confusion is understandable.

No doubt, all of you have received a University education. As that is the case, you are *aware* of the revolutionary theories of Marx and Lenin. Bear in mind that there is a big difference between being aware of those theories, and having a proper scientific *understanding* of those theories.

My point is that the Universities teach only the bourgeois *distortions* of the theories of Marx and Lenin. That is very likely the reason so many middle class people, who consider themselves to be Socialists, are of the opinion that Socialism is a good idea, but simply not possible! They have learned this in University!

It was Marx who conducted a thorough, *Scientific* examination of capitalism, and *proved* that it would, of *necessity*, give rise to Socialism!

Here is the way Marx phrased it: "And now as to myself, no credit is due to me for discovering the existence of classes in modern society, nor yet the struggle between them. Long before me, bourgeois historians had described the historical development of this class struggle, and bourgeois economists the economic anatomy of the classes. What I did that was new was to prove: 1) that the *existence of classes* is only bound up with *particular historical phases in the development of production;* 2) that the class struggle *necessarily* leads to the *Dictatorship of the Proletariat*; 3) that this Dictatorship itself only constitutes the transition to the *abolition of all classes and to a classes society*".

Take note, Marx made it clear that, under capitalism, the "class struggle *necessarily* leads to the *Dictatorship of the Proletariat!*

With that in mind, may I suggest that you once again read the most important revolutionary works of Marx and Lenin, but with an open mind. Disregard the bourgeois distortions! No doubt, you will learn, perhaps to your surprise, that the Scientific Socialists, the Communists, are correct, that in fact it is true, that capitalism *necessarily* leads to Socialism, in the form of the *Dictatorship of the Proletariat!*

This stands in stark contrast to the University distortions of those theories. They teach the utopian Socialist viewpoint that, while Socialism may be a fine idea, it is simply not possible. Such is hardly the case!

Once you have a true understanding of those revolutionary theories, feel free to bring that understanding to the working class.

The importance of this must be stressed. In fact, as Lenin stated, "Engels recognizes *not two* forms of the great struggle Social Democracy is conducting (political and economic), as is the fashion among us, *but three, adding to the first two* the theoretical struggle." (italics by Lenin, while at that time Marxism was referred to as Social Democracy)

I mention this in order to stress the importance of raising the level of awareness of the proletariat.

Prepare the Working Class For Council Power and the Dictatorship of the Proletariat

In particular, the working class must be made aware of the fact that the existing state apparatus must be *smashed*, at the time of the Revolution, and replaced with a new, Proletarian state apparatus, in order to crush the capitalists, as they make every effort, after the Revolution, to restore their "paradise lost". This state apparatus is referred to as the Dictatorship of the Proletariat. We will know we are being successful when that expression becomes common place.

Rest assured, as long as you voluntarily join the revolutionary motion, in the fight for Scientific Socialism, your past will not be held against you. On the contrary, your experience in the service of the billionaires, can prove to be most valuable. Feel free to share that experience with the working people. This will serve to drive home the fact that classes exist, and that our differences are antagonistic. The proletariat must be persuaded that the capitalists, the billionaires, must be overthrown and crushed.

In the interests of combining theory and practice, it is perhaps best to become involved with the various Councils, or Soviets, which tend to take shape during a time of revolutionary motion. No doubt some of these Councils are involved in locating these huge animals. As I have gone into this in other articles, there is no need to repeat it here.

As well, get together with other middle class Marxist intellectuals, as well as advanced workers, and take part in the formation of a proper Communist Party, one which calls for the Dictatorship of the Proletariat. Discretely! There is an urgent need for such a Party.

The fact of the matter is that, after the successful Socialist Revolution, under the Dictatorship of the Proletariat, your services will be required and rewarded. Socialism requires professional people, managers, engineers and scientists, in all fields. For the most part, you have performed a fine job for the capitalists. Under Socialism, you will do an even better job, for the proletariat. After all, the atmosphere at the workplace will be far more relaxed. You will not have to worry about office politics! We do not play such games!

Bear in mind that middle class intellectuals, including Socialists, are well aware that revolutions happen, and on a regular basis. They are familiar with the Russian February Revolution of 1917, which removed the Emperor from the throne. This gave rise to the democratic republic of the Russian capitalists. That was as far as the working people could go! They could not go further, to Socialism, because they were not aware of the revolutionary theories of Marx!

It was only *after* Lenin returned from exile, in April of 1917, and provided them with the proper revolutionary theory, which included the Dictatorship of the Proletariat, that the common people, the workers and poor peasants, could overthrow the capitalists, and establish a proper Socialist society.

In much the same manner, the forthcoming American Revolution can also go only so far, *unless* the working class is aware that the existing state apparatus must be *destroyed*, and then *replaced*, with the Dictatorship of the Proletariat!

The alternative is to overthrow one set of capitalist rulers, and merely replace it with another set! Not a vast improvement!

The American Revolution could break out at any day. There is no time to lose.

Advice To All Professionals

With that in mind, I can only offer a word of friendly advice to the scientists and government officials, those who persist in propagating those ridiculous theories. Feel free to "come clean", to admit that it is all a "pack of lies". After all, any day now, the truth will be revealed. Better to "cut your losses", to join the working class in overthrowing the billionaires.

The middle class is currently in the process of being *destroyed*, facing financial ruin. All are being forced into the ranks of the working class, the proletariat.

As for those who are skeptical, feel free to face the fact that the billionaires are in charge, and they have recently *laid down the law!* They "tipped their hand", so to speak! They announced that a mere *eight banks*, as well as *five businesses*, are *Too Big To Fail!*

Perhaps they have overlooked one little detail! *Every coin has two sides!* The other side of that coin is that there are *thousands of banks, and tens of thousands of businesses*, which are *Too Small To Succeed!* All of those banks and businesses are about to *fail*! Even General Motors is *Too Small To Succeed!*

As all of these countless banks and businesses fail, the few that are Too Big To Fail, will pick up their "assets", and the billionaires will become ever more wealthy! Their goal of becoming *Trillionaires* is within reach!

There are those who object that the billionaires would never do that, because it would ruin the country! Drive it back into the Middle Ages! True! Yet that is not the *goal* of the billionaires! Do not give them too much credit! They are not trying to drive this country into the ground! They simply *do not care!* The success- or failure!- of the country, is a matter of complete indifference to them! They are entirely focused on themselves! They must be stopped! But how? This brings us to our next topic.

The Industrial Revolution and the Creation of the Revolutionary Proletariat

All previous civilizations have risen to a peak, and then fallen into decline, eventually collapsing altogether. Yet our civilization is *not* destined to go the way of all previous civilizations, because our civilization has experienced the one thing which no previous civilization has experienced. An industrial revolution!

The importance of this revolution cannot be overstated. It was the greatest thing to happen to humanity, since the domestication of plants and animals! It not only revolutionized production, it created *two revolutionary classes!*

At *first*, both newly created classes were revolutionary! That which the new revolutionary class of capitalists created, is sure to remove all doubt! Magnificent creations!

All of this took place at a time of early, competitive capitalism.

Then, around the beginning of the twentieth century, capitalism reached the stage of monopoly, referred to as imperialism. At that point, it became completely reactionary!

Lenin conducted a thorough study of monopoly capitalism, and wrote a book, titled Imperialism, the Highest Stage of Capitalism. This is one more work of Lenin which I highly recommend!

The point is that the capitalist class is now completely reactionary, and is driving our civilization into the ground. Yet the proletariat is a completely revolutionary class, and is about to prevent the billionaires from succeeding in this!

Now the problem is one of making them aware of this! The middle class people can be of assistance in raising the level of awareness of the proletariat!

Your education and experience can be put to good use. Feel free to share with the working people, the revolutionary theories of Marx and Lenin. Let them know what you have gone through, in your years of service to the billionaires. Tell people how you degraded yourself, spreading lies, in the interests of maintaining your career. Your past will not be held against you.

Bear in mind that the alternative is to oppose the revolution, to continue to defend the billionaires. In that case, you can expect to become a target of the revolution. The working class cannot force you, or anyone else, to do the right thing. Yet under the Dictatorship of the Proletariat, we can make you wish you had!

At the time of the revolution, your capitalist masters will not be able to protect you! Nor do they have any desire to do so! The billionaires care only about themselves, and their immediate families! They will not hesitate to throw you under the bus!

Choose wisely! Otherwise, do not be surprised if, after the Revolution, you end up in the same remote work camp as that of your former lords and masters, the billionaires. You may then find yourself performing the same manual labour, which could well involve work with pick and shovel. You have been warned!

The fact of the matter is that the forthcoming American Revolution can also go only so far, *unless* the working class is aware that the existing state apparatus must be *destroyed*, and *replaced*, with the Dictatorship of the Proletariat!

The alternative is to overthrow one set of capitalist rulers, and merely replace it with another set! Not a vast improvement!

The Revolution could break out at any day. There is no time to lose. For that reason, our slogan must be:

Prepare For Council Power and the Dictatorship of the Proletariat!

PART 6

BECOMING ACTIVE

Workers: Assist In Locating the Big Three!

All common people, workers and family farmers, are well aware of numerous myths and legends, concerning various animals, many of which are huge and terrifying. This is because of the stories, which are passed down, from one generation to the next. They find these stories fascinating, if only because they are based upon animals which exist.

By contrast, the scientists are also well aware of these legends, and dismiss them out of hand. They refer to them as "Old wives tales", and "pure nonsense". In this, the scientists are completely mistaken. But then, it is in their best interests to disregard such beliefs.

Remarkably enough, most of these legends are based upon animals which fall into three categories. These include those that fly, those that swim, and the third is that of giant people.

All of these animals are very much alive, and can be located here in North America. They are absolutely not extinct, and with the help of working people, we can quite easily prove this.

As for the skeptical, astute reader, who is wondering just how this is possible, may I refer you to a speech given by Lenin, in October of 1920, three years after the Great October Socialist Revolution, to a gathering of the Young Communist League: "It was the declared aim of the old type of school to produce men with an all round education, to teach the sciences in general. We know that this was utterly false, since the whole of society was based and maintained on the division of people into classes, into exploiters and oppressed. Since they were thoroughly imbued with the class spirit, the old schools naturally gave knowledge only to the children of the bourgeoisie. Every word was falsified in the interests of the bourgeoisie. In these schools the younger generation of workers and peasants were not so much educated as drilled in such a way as to be useful servants of the bourgeoisie, able to create profits for it without disturbing its peace and leisure."

Of course Lenin was referring to the bourgeois schools, the schools under capitalism, those controlled by the billionaires. Those are the very schools with which we are currently forced to attend.

As a result of this, those who graduate from these schools, including those who have earned degrees in science, are truly "servants of the bourgeoisie", those who know how to "create profits", without "disturbing its peace and leisure".

Without doubt, the act of locating these three particular species of animals, will certainly "disturb" the "peace and leisure" of the bourgeoisie, the billionaires. All the more reason to find them!

Dragon Or Thunder Bird

The animal which has given rise to the most legends, is the flying reptiles, technically referred to as pterosaurs, more commonly referred to as pterodactyls. In most parts of the world, such as Asia, Africa and Europe, they are referred to as Dragons. Here in North America, there are numerous local names. The more common local names include Thunder Bird, Devil Bird, Satan Bird, Demon Bird and Jersey Devil. These animals are nocturnal, coming out of the caves, in the mountains, mainly after sundown, and hunting in darkness.

They prey upon people, as well as livestock. It is only natural that common people should get the idea that these animals are demons!

Of course, they are not demons, but the beliefs of all common people must be respected. It is also absolutely essential that we prove that they exist, if for no other reason than to give warning to people.

The fact is that they are predators, carnivores, and as such, have a very keen sense of smell. Further, they are far more likely to attack, at the smell of blood. For that reason, they consistently prey upon girls of childbearing age. They also prey upon children, as they are easier to pick up and carry away. Of course, on occasion, they also prey upon men.

This brings me to the best method of proving their existence.

Bear in mind that they are famous for releasing a "cloud of smoke", so called, which is not smoke at all, but it is toxic. Poison gas. In fact, it is so strong, it is capable of killing horses and cattle. We know this, for a fact, because that is precisely the manner is which they kill those animals. Then they rip the flesh from the face, as well as the genitals, and frequently tear out the large intestine.

Then in the morning, the owner of these animals stumbles upon these carcasses, furious and bewildered. Such owners generally call the police, which does no good, as it is not a police matter. After all, no laws have been broken. Humans were not involved in these killings. There is no law against wild animals killing livestock. If anything, it is a matter for the game warden. Yet it is rare that the game warden is even notified.

This brings me to the method we can use to prove that these predators exist. First, as soon as possible after the animal is killed, or at least within twenty four hours, draw out a sample of blood from the carcass. Send this sample to the lab, to be analyzed. The technicians in the lab will in turn determine the poison gas that was used to kill the animal. I am certainly curious.

Second, at the same time, swab around the wound sites, and send those swabs also to the lab, for a DNA analysis. No doubt the lab will determine that the wounds were inflicted by a reptile, one which is not known to science.

The third thing which has to be done is a bit more complicated, and requires a little explanation.

Close to the village in which I live, that of Tsay Keh Dene, there is a range of mountains, which the locals refer to as Buffalo Head. These mountains are quite distinctive, in that many of them have a very steep, or vertical face, commonly referred to as a cliff, or bluff, while the top of the mountain is horizontal, or flat. I refer to these as "perpendicular mountains". It is such mountains that contain the nesting ground of the pterosaurs.

Thirty years ago, a logging road was built into that area, referred to as the Ten Thousand Road. The caves in those mountains, which are the nesting grounds of the pterosaurs, open up onto that Road. Now it is a little matter of accessing those caves and placing cameras near the openings.

After the area was logged out, the Road was "deactivated". A backhoe tore up the Road, and the two bridges were taken out, so that no one could go into that area. This is standard procedure.

However, now that same Road has been opened up, and the area is currently being logged again. One bridge has been put back in place, and the other bridge is scheduled to be installed next summer. The nesting ground of the pterosaurs is just beyond that second creek.

In order to access the nesting ground of the pterosaurs, it is necessary to get the permits to place a bridge over that creek, and open up the remainder of that Road, up to the nesting ground.

Of course that requires government permits, as well as money, in order to hire the proper equipment. Yet once we establish the fact that a great deal of livestock is being killed by poison gas, released from an animal which is not known to science, then no doubt sufficient pressure can be brought to bear upon the officials. As that is the case, in due time, we can expect them to become sweetly reasonable.

Sasquatch Or Bigfoot

The second animal is also the source of a great many legends. It is nothing other than a separate species of human, referred to as Gigantopithecus, commonly called Sasquatch or Bigfoot. I refer to them as Giants, as they are huge, and the name is not derogatory. It is essential that we prove they exist, in order to pass laws to protect them. They have the right to live their lives as they see fit. We have no right to interfere with them. We certainly have no right to hunt them!

These Giants are nomads, members of a hunting-gathering society, so that they are frequently on the move, avoiding us. They know what a vicious bunch we are! Yet they also know that the Indigenous People are more tolerant. For that reason, on the Pacific Coast, the Giants frequently go onto the beaches, on the Reserves, after sundown. The Indigenous People respect the Giants, just as the Giants respect the Indigenous People. That is the way it should be!

That is also the place we can establish contact with the Giants. Rather than hunt them, we want to attract them. Allow them to come to us!

First is the not so little matter of securing the cooperation of the Elders, who live on the Reserves. We must explain to them that we merely want to meet the Giants, in order to prove that they exist. Assuming we can get the cooperation, of the Elders, then it is a matter of putting out gifts for the Giants, on the beach, before sundown. Bear in mind that as they are human, they love the same things we love. That includes food such as vegetables, fruit, meat, fish and cheese, raw as well as cooked. Also, may I suggest steel mirrors, as they too want to know what they look like. Then there are cosmetics, in order to enhance their appearance. No doubt, they have baskets, but as they tend to be rather flimsy, more sturdy burlap bags will be appreciated.

I can only stress the fact that we want to attract them. In due time, they will respond to our advances. Give them time, as they have no reason to trust us! Yet they trust the Indigenous People, so perhaps the breakthrough will take place somewhat sooner, rather than later.

Ogopogo Or the Loch Ness Monster

The third item on our "hit parade" is also a source of numerous legends. In North America, it is most commonly referred to as Ogopogo, as it is frequently seen in Okanagan Lake. No doubt, that is a rather clever play on words. Mind you, the animal in Lake Champlain is called Champie. In Lake Erie, that same animal is called Bessie, probably short for basilosaurus. Also very clever!

Then there is the animal in Loch Ness, called Nessie, if you can believe it. Very likely the same animal, or animals, although that remains to be determined.

As I have just documented this in the previous article, there is no point to repeat it here. Suffice it to say that a simple trail camera is all that is needed.

These are quite simple tasks, yet also revolutionary. The billionaires, the bourgeoisie, have to be overthrown and crushed, under the Dictatorship of the Proletariat. In order to accomplish that, the working class, the proletariat, has to become active. This is one manner in which workers can make a difference. In fact, they can take part in a number of major scientific break throughs.

It may help to think of this as preparation for the Dictatorship of the Proletariat, because that is precisely the case. After the revolution, workers will be placed in positions of authority, within the new socialist government. Any training workers receive now, will prove to be of great value, after the revolution.

My hope is that recently formed Councils (Soviets), will come together and cooperate in proving the existence of these magnificent animals. These

Councils can also form the nucleus of revolutionary bodies. Success will no doubt attract others, so that the Councils are bound to grow. Let the slogan be:

Prepare For Council Power and the Dictatorship of the Proletariat!

PART 7

SPIRITUAL POWER

Introduction

As is well known, the Communist Manifesto was written in 1847. At that time, Marx and Engels were well aware that after the successful socialist revolution, it would be necessary to have "the proletariat organized as the ruling class".

Of course, they had no idea of the precise form this organization would take, as they had no experience of any such successful revolution. That changed dramatically with the appearance of the Paris Commune, in the spring of 1871.

Although the Commune was brief, lasting a mere few weeks before it was crushed, with great brutality, Marx subjected it to a most careful analysis in his book, The Civil War In France.

Perhaps the most important lesson he drew from this French Revolution, was the fact that the workers cannot merely lay hold of the existing state machine and use it for their own purposes. Instead, it has to be smashed, and replaced with a new state apparatus, in the form of the Dictatorship of the Proletariat.

This fact is well known to the social chauvinists, the revisionists. Which is not to say that they have embraced this revolutionary Marxist theory. On the contrary, they avoid any mention of this, unless of course it involves distorting the theory. After all, they have their own agenda, which includes taking over the state apparatus, at the time of the revolution, and setting themselves up as the new rulers. The destruction of the existing state machine is the last thing the social chauvinists want to see!

I mention this because it is of such vital importance. Marx went on to say more: "Having once got rid of the standing army and the police, the physical force elements of the old government, the Commune was anxious to break the *spiritual force of repression*, the 'Parson Power'".

My point is that the bourgeois state apparatus consists of more than just the "physical force elements of repression", that of police, standing army National Guard, prisons and various "correctional institutions".

It also consists of numerous "spiritual forces of repression", and that too has to be smashed, at the time of the Revolution, *as it is part of the state apparatus!*

In the case of the Paris Commune, Marx referred to the spiritual force of repression of the clergy, as "Parson Power". It may help to think of these "spiritual forces of repression" as "invisible chains", and these chains have to be broken!

Various Spiritual Forces of Repression

These "spiritual forces of repression" assume many forms, in different parts of the world.

In China, at the time of the Revolution, the peasants were rising up and overthrowing the landlords. As they had been crushed and exploited by the landlords for centuries, they lived in mortal dread of those people. Properly so, I might add! For that reason, it was not enough to separate the landlords from their wealth and property. The spiritual power of the landlords had to be broken!

For that reason, the peasants also placed "dunce caps" on the landlords and paraded them through the streets, mocking and humiliating them. It was in this way that the "spiritual force of repression", the "invisible chains" of the landlords, was destroyed!

If we compare this to the American Revolutionary War of 1776, then it is safe to say that the Chinese landlords "got off easy". In the American Revolution, the British Nationalists, the "Tories", those who were crushing and exploiting the American Colonials, were first coated with tar and feathers, before being paraded though the streets. Lo and behold, the spiritual power, the "invisible chains" of the Tories, was broken!

Of course, such behaviour is considered to be socially unacceptable, under non revolutionary situations. Yet in times of revolution, the polite customs of civilized society tend to "fall by the wayside".

The fact remains that all "spiritual forces of repression", all "invisible chains", have to be smashed. In China, it was the landlords who terrorized the peasants. In the American colonies, it was the Tories whom the colonials dreaded. Both were dealt with, rather harshly, by the common people, the members of the public. In both cases, the result was the same. The "spiritual power," the "invisible chains", of the oppressors, was broken.

Spiritual Power of Mobsters

The current situation in North America is similar, in that there are numerous mobsters, or members of "Organized Crime", to use the politically correct expression, who make a career of terrorizing and exploiting the working people.

One of their favourite methods consists in extorting "protection money" from small business owners, a form of "insurance", to keep their lives and property safe. Those who do not "pay up", generally live to regret this, but by no means all. Some do not live to regret it!

No doubt, at the time of the approaching American Revolution, the "spiritual force" exerted by these mobsters will also be broken. It is very likely that working class courts, which may consist of three judges, Tribunals, may order that these mobsters first be publicly humiliated, before serving their sentence.

This humiliation can take different forms. Different areas will no doubt resort to different forms of humiliation. These could include such things as forcing the mobsters to wear ridiculous clothes, such as clown costumes, or women's garments, shaving their heads and wearing dunce caps. Or they could be coated with tar and feathers!

Then they will be forced to walk through the streets, of the neighbourhoods of the people whom they have previously exploited and terrorized. The residents will then be allowed to pelt those mobsters with garbage!

It is in this way -and only in this way!- that the spiritual power of the mobsters can be broken! It is only *after* the mobsters have been publicly humiliated, *after* the spiritual power of those mobsters has been broken, that those people will serve their sentences!

At the time that this is taking place, we can expect the "bleeding heart liberals" to "cry crocodile tears"! They will bemoan the fact that these "poor unfortunates", these "citizens", have been denied their democratic rights!

That is the whole idea! That is the reason it is called the *Dictatorship of the Proletariat!* Under this *Dictatorship*, it is *not* only the capitalists, the billionaires, who will have *no rights!* The thieves and killers also, will have *no rights!* The sex offenders will have *no rights!* The dope dealers will have *no rights!* In fact *all* of the human low life will have *no rights!* That is the reason it is called a *Dictatorship!* The common people, the working people, the *victims* of those human predators, will have *all the rights!*

We can expect others, those who are more advanced, to respond that this public humiliation may be terrible, but it is necessary. That is closer to the truth, but still incorrect. In fact, this public humiliation, of those who have devoted their lives to exploiting and terrorizing the working people, is both *necessary and wonderful!* The spiritual power of the mobsters must be broken! This is only possible through public humiliation!

It is not at all terrible! It is the manner in which the working class breaks the invisible chains of spiritual power! Rejoice! The working class is liberating itself!

Yet there are other forms of spiritual power, some of which are far more subtle, but just as real. As they are more subtle, I would argue that they are more dangerous!

Professor Power

Of course, I am referring to the spiritual form of repression exercised by the intellectuals, especially those who are responsible for the school text books. Many of them, if not all, have been written by Professors, at various Universities. For that reason, I refer to them as "University Professors", and as they possess considerable spiritual power, I refer to this as "Professor Power".

The source of this spiritual power comes not from the terror inspired by the threat of brute force, as per the mobsters, still less from the threat of everlasting damnation, as per the Clergy, but from the *unquestioning* respect, which is *demanded* by these intellectuals.

The University Professors have established themselves as a force which is above reproach! They have set themselves up "on a pedestal", and *certain* of the scientific theories which they have endorsed *are not to be challenged!* *"Sacred" Scientific Theories!*

They have managed to gather to themselves a considerable amount of spiritual power, "Professor Power", through their control of the educational system. Perhaps a few examples will serve to illustrate this.

All scientists are agreed that there was a mass extinction of dinosaurs, sixty five million years ago. As far as the theory of the mass extinction of dinosaurs is concerned, that is about the only point upon which they are agreed! Certain science books maintain that dinosaurs were "active, intelligent, warm blooded animals". Other science books insist that dinosaurs were land dwelling reptiles. Still others insist that this included flying reptiles. Other books of science insist that swimming reptiles were also dinosaurs.

The scientists are also deeply divided over the cause of this extinction. The most widely held belief is that of a "killer rock from outer space". That has captured the imagination of countless common people! That is more popular than the theory of continental drift, "galloping continents", which carried diseases in their wake!

Let us not forget the theory that the dinosaurs grew too big, so that they were unable to have sex, thus could not reproduce!

That last one has got to be my favourite! To think that such nonsense is put forward in the name of science! It is nothing short of an embarrassment! But then perhaps the scientist who dreamed up this "theory" is on drugs, some real good stuff. We should all be so lucky! No need to be greedy! Feel free to share!

Of course the only reason that the scientists cannot give a rational explanation for the mass extinction of dinosaurs, is because there is none! There was *no* mass extinction of dinosaurs! Most of the animals which have been classified as dinosaurs, were birds! Others were reptiles! Yet there was evolution, so that over the course of many millions of years, many species of birds have evolved, so that new species came into existence.

Then too, there were the reptiles, some of whom have also been classified as dinosaurs. They too have evolved, but as they are cold blooded animals, they evolve much more slowly than birds. They remain largely unchanged.

As well, all scientists are also agreed that we are the one and only species of human still walking the earth. All eye witness reports of another species of human, that of Gigantopithecus, commonly referred to as Sasquatch, Bigfoot or Giants, are calmly and quietly disregarded!

Then there is the theory of the "mass extinction of mega fauna", in which huge animals, including the woolly mammoth and sabre toothed cat, dropped dead, at the end of the last ice age, because they "could not survive climate change". The fact that these animals survived three ice ages, which included several changes in climate, is also ignored!

Of all the denials, by the scientists, perhaps the most common place, is that of the existence of "UFO's", or "Unidentified Flying Objects". Regardless of the number of reliable eye witness accounts, as well as videos, the scientists still maintain that they do not exist! In terms of being stubborn, they put mules to shame!

In fact, this "glow in the night sky" is nothing other than the bioluminescent lights emitted from the torsos of the male pterosaurs, also known as pterodactyls. Yet as the scientists still maintain that those flying reptiles went extinct, many millions of years ago, they disregard all such sightings! Their capacity for self delusion is an absolute marvel!

This brings me to another theory upon which all scientists are agreed, and that is the fact that there was a time, fifty million years ago, when whales walked on land. True! But then they are still walking on land! Walking whales! Twenty meters long, or sixty five feet, so that they are quite hard to miss. Yet the scientists have managed! I have to give them credit!

It is not just whales that still have legs, and still walk on land! Also ichthyosaurs! In fact, as the various legends concerning the "Loch Ness Monster", include the detail that "Nessie has fins", it is more likely that "Nessie" is a reference to the ichthyosaur. After all, these animals have fins on their back. The walking whale, basilosaurus, has no such fin.

That is a mere selection of the various "scientific fairy tales" which are contained in the school text books. Students of all ages, from grammar school to university level, are required to memorize this nonsense, as a means of earning a diploma or degree. That is sad.

It is also "spiritual power". This power, which I refer to as "Professor Power", is not to be underestimated!

Spiritual Power In Education

The common people have been told, all their lives, that the key to success is through education. There is some truth to this, as in a highly industrialized country, illiteracy is a severe handicap. Yet at the same time, the school text books have been written by the University Professors. The one and only way a person can pass all the courses and earn diploma or degree, and in turn become successful, is by first memorizing all the information which is contained in those text books.

This information is certainly not limited to science. The history books are also filled with distortions and "fairy tales", out right lies. The point being that the "key to success" lies in not "rocking the boat", *not* pointing out the lies and hypocrisy contained within these text books, all of which have been written by the University Professors. That is real power! Spiritual power! Professor Power!

As for those who suspect that perhaps I am overstating the matter, may I refer you to a speech given by Lenin in August of 1918, less than one year after the successful Russian Socialist Revolution: "The working people are thirsting for knowledge because they need it to win. Nine out of ten of the working people have realized that knowledge is a weapon in their struggle for emancipation, that their failures are due to lack of education...they see how indispensable education is for the victorious conclusion of their struggle."

Under the current political situation, that of capitalism, Lenin pointed out that the schools are nothing but "an instrument of the class rule of the bourgeoisie". Their purpose is to provide the capitalists with "obedient lackeys" and "able workers". Those who properly commit to memory all the distortions and lies, contained within the school text books, are able to earn a degree and serve as "obedient lackeys". They are among the most devoted servants of the capitalists, the billionaires.

That in no way changes the fact that the working people are "thirsting for knowledge", and believe that the school text books can provide them with that knowledge. Who can blame them?

By the same token, the University Professors are well aware that it is in their best interests to perpetuate the same old myths, the lies of the capitalists.

They have chosen to embrace a life of hypocrisy, in the service of the capitalists, and are reasonably well paid for this.

To each his own.

Necessity of a Scientific Revolution

Working people have got to learn, from their own *experience*, that the University Professors are in the service of the capitalists. They must be "knocked off their pedestal".

This is *not* to say that we must wait until after the revolution, after we overthrow the capitalists, after we smash the existing state machine and establish the Dictatorship of the Proletariat.

There is no need to wait! We can take action *now*! By proving that at least a few of their cherished theories are mistaken, we can help to *empower* the working class, to break the invisible chains, the *spiritual power* of the University Professors! At the same time, we are *preparing* the workers for the approaching revolution, providing them with valuable experience for the subsequent *Dictatorship of the Proletariat!*

It may help to think of this as a "double play", or as "killing two birds with one stone", to put it is popular terms. After all, it is my belief that the most important thing now, the "key link", consists in preparing the working class for Soviet (Council) Power, and the Dictatorship of the Proletariat.

There are various aspects to this preparation. On the one hand, all workers have to be made aware of themselves as a class, with their own class interests, as opposed to the interests of the capitalists, the billionaires, the bourgeoisie. They also have to become aware of the necessity of smashing the existing state machine, at the time of the Revolution, and replacing it with a new, working class state apparatus, in the form of the Dictatorship of the Proletariat. A careful reading of State and Revolution, by Lenin, will go a long way towards this awareness.

On the other hand, a little practical experience goes a long way! It is not enough to "talk the talk"! We have to "walk the walk"!

For that reason, I recommend that all working people get involved with the Councils that are appearing, all across the country. Part of the work of those Councils can involve proving the existence of these huge animals. Animals which the scientists swear are extinct!

In this way, the self confidence of the working people will be raised. They will see for themselves that the scientists are largely a pack of liars! Hypocrites! This will help to break the spiritual power of the scientists, the "invisible chains", the "Professor Power", of the scientists! They will be exposed as the loyal and devoted servants of the billionaires, which they are!

It would be best if certain intellectual members of the middle class helps with this, preferably as part of the work of various Councils. At the same time, those intellectuals could evaluate the actions taken by the workers, with a view to placing the most dedicated workers in key positions, at the time of the insurrection.

After the revolution, those same advanced workers will be placed in key positions of authority, under the Dictatorship of the Proletariat. Any training and experience they gain now, will prove to be most valuable! All part of preparing for Council Power and the Dictatorship of the Proletariat!

Allow me to draw a clear distinction between us, Communists, and anarchists, those who believe in no government. This is to say that after the revolution, the billionaires will make every effort to restore their "paradise lost"! That is the reason we still need a state apparatus, in order to crush their desperate and determined resistance! That state apparatus is referred to as the Dictatorship of the Proletariat! Within that new state apparatus, certain key positions of authority will have to be filled, by advanced workers. Any training they receive now will prove to be most valuable!

As mentioned in a previous article, all scientists maintain that basilosaurus, the "walking whale", is extinct. It most certainly is *not* extinct. Neither is the ichthyosaur! In North America, they are most commonly referred to as Ogopogo, and are located within Okanagan Lake, as well as many other large lakes.

They are most likely also located in Europe. Perhaps the most famous one being "Nessie", the Loch Ness Monster. My niece, who is Italian, tells me it is also in Italy. As these animals come out of the water to graze, after sundown, proving their existence should not be terribly difficult. The only thing required is an inexpensive, motion activated, trail camera.

The working people who take part in the location of these huge animals, will be taking part in a major scientific breakthrough. It is not just the fresh water swimming mammals that can be easily proven to exist. The act of proving they exist, in defiance of all scientists, can have no other effect than that of "knocking the scientists off their pedestal", of "breaking the invisible chains" of overcoming the spiritual power of the professors, the "Professor Power".

Although the scientists will no doubt find this to be unpleasant, perhaps they can take some comfort in the fact that the alternative, that of being "tarred and feathered", as were their predecessors, is so much worse!

Then again, at the time of the revolution, the working people will decide upon a course of action, focused upon their class enemies. All professional people would be well advised to bear this in mind. As yet, we have no idea of the precise form of that action. We just know that it will not be pleasant!

In the broadest sense of the term, this "Professor Power" can be extended to all people who have degrees from Universities, and work in various fields, whether in government, education or industry. All are well aware that to ask "awkward question", to "rock the boat", is to risk "career suicide". This can result in being fired, and worse, being "black balled", so that they are never again allowed to work in their chosen field. That is real power! Spiritual power! Professor Power!

Challenge All Scientific Theories

On a related matter, perhaps a little explanation is in order. I have my own personal religious beliefs, which I prefer to keep to myself. Properly so, I might add. I certainly have no quarrel with those who have other religious beliefs, and in particular, not with any clergy! It is important that all people, scientists as well as common people, respect the personal beliefs of each other!

That stands in stark contrast to scientific theories, which are nothing more than scientific beliefs, and are *meant to be challenged!* Under no circumstances should this challenging of scientific theories be taken as a personal attack! Yet all too often, that is precisely the case!

As for the accusation that I have a prejudice against scientists, nothing can be further from the truth! In fact, I take inspiration from the pioneers of science, such as Kepler, Copernicus and Galileo. Those men risked their lives in challenging the ecocentric theory of the time, assuming that is the correct term. That was the belief that the earth was at the centre of the universe. If they had been caught, they would have been executed!

It was Newton who collected the work of those scientific pioneers and managed to put it all together. This resulted in his three laws of motion. As Newton put it, he merely "stood upon the shoulders of giants".

As a result of this, the "ice was broken", so to speak, and other scientists dared to speak up. In particular, Darwin bravely suggested the Theory of Evolution by Natural Selection. To this day, that has managed to rouse the wrath of numerous "fundamental" people, of various religious beliefs. They maintain that it fails to account for Creation!

To this I can once again respond that we should all respect the *personal beliefs* of all others, scientists as well as common people. That includes the belief in Creation! At the same time, all scientists *must* put aside their personal beliefs, when evaluating scientific theories!

All working people, scientists as well as common people, have a common enemy! That common enemy is the monopoly capitalists, the billionaires! They must be overthrown! They must then be crushed, under the Dictatorship of the Proletariat!

This requires a united front! We cannot afford to fight among ourselves! We must put aside our personal differences, and work towards that common goal!

This brings us to a little problem which must be faced. That problem is one of dinosaurs.

Many years ago, numerous scientists were faced with the puzzle of the bones of huge animals, which must have lived many years ago. As the bible makes no mention of such animals, certain scientists were able to muster the courage to suggest that the bible was mistaken!

The problem was that at that time, by which I mean the time in which Darwin worked, it was thought that all the answers were located in the bible. There are a great many people who are still of that opinion. For my part, I respect those beliefs. But as perviously mentioned, I can only stress my opinion that our personal beliefs should be kept separate from the scientific theories.

So the scientists of that day, the earliest palaeontologists, did the best they could and named these pre historic animals "dinosaurs". The choice of names was unfortunate, as it means "terrible lizard", and we now know that most of these animals were birds. That in no way changes the fact that the word "dinosaur" is now a house hold name. It is just one of those things with which we have to live.

Yet the scientific breakthrough had been made, and numerous others followed. The Egyptian Rosetta Stone was deciphered, and that made possible the translation of Egyptian hieroglyphs. In North America, Morgan was able to determine the social structure of primitive society. That resolved the earliest mysteries of Roman and Greek civilizations. In South America, the ruins of various civilizations were discovered. Penicillin and various antibiotics were discovered. The transistor was invented. Let us not forget the aircraft and the internal combustion engine. All of these things, and many others, are commonly referred to as "progress".

So how is it that a great deal of scientific gobbledegook has found its way into the school text books? It clearly suggests that the "rot has set in", that we are making progress "in reverse", that "we are going backwards". And so we are!

Decline of Our Civilization

The fact is that over the last few thousand years, numerous civilizations have come into existence, risen to a peak, prospered, and then fell into decline. They left behind the ruins, so as to remove all doubt!

Our civilization has also passed its peak, and is now in decline. That is not acceptable. Yet there is a difference between our civilization and all previous civilizations. Our civilization was the first to experience an industrial revolution. As a result of that revolution, two new classes were created, that of capitalists, bourgeoisie, and workers, proletarians. Now it is up to us, the working class, the proletariat, to reverse this decline. All previous civilizations could not prevent their own decline, because they did not have a progressive class of proletarians. We are the first civilization that is able to reverse our own decline! Now that calls for a little explanation.

The fact is that we live under a state of monopoly capitalism, referred to as imperialism. All imperialists have one thing in common: They are all reactionaries! There is nothing progressive about them! They care only about themselves! In their quest for an ever greater profit, they are driving our civilization into the ground. They must be destroyed! That is where the proletariat comes into play!

This is not to say that the capitalists, the billionaires, are consciously determined to destroy our civilization. They are not! To suggest that is to give them too much credit. Nor are they trying to save our civilization. They simply do not care, one way or the other. The subject is a matter of complete indifference to them. They have given it no thought. They are focused on making a profit, becoming ever more wealthy, preferably a Trillionaire. Nothing else matters!

With that in mind, may I suggest that everyone face the fact that we live in a class society. As that is the case, either we serve the working class, the proletariat, or we serve the capitalist class, the billionaires, the bourgeoisie. There is no middle ground! Anyone who serves the billionaires, and that includes the University Professors and scientists, can and will find themselves a target of the revolution. Fair warning! That is just the way it is.

In fact, the professional people, whether Professors, scientists, engineers, teachers or administrators, have nothing to lose and everything to gain, by the revolution. Under capitalism, their jobs are never entirely secure. Regardless of how well they perform their jobs, they can be fired at any time, for any reason, or for no reason. The capitalists may choose to fire someone because they can!

By way of contrast, after the revolution, under the Dictatorship of the Proletariat, these professional people will play a key role in the new socialist society. They are currently serving the capitalists well. Soon, after the revolution, they will serve the working class just as well, probably much better. They will be treated with the respect they deserve, and paid quite handsomely. They will not have to commit to memory all the lies of the capitalists. They will not be subjected to the petty bickering and backstabbing, which is characteristic of all capitalist establishments. The work place atmosphere will be far more relaxed. Those who perform an exceptional job, can expect to be promoted.

Part of that job performance will involve training certain workers, those who are more advanced, in such technical work. The "cut throat" days of "every man for himself" has no place under socialism! Instead, the competition will be to perform to the highest standard.

The factories, for example, which perform to the highest level, will serve as "model factories", in order to inspire the workers in other factories to perform just as well.

No doubt, there will be a period of adjustment, as people become accustomed to life under socialism. For the vast majority of people, this will be voluntary, as they embrace the improved conditions. Then again, for a significant minority, a certain amount of coercion will be required.

In particular, the monopoly capitalists, the billionaires, will fight this "tooth and nail". This we can practically guarantee, as their standard of living will "nose dive". They will pine for the "good old days", when servants catered to their every whim, fleets of vehicles were at their disposal, clothes were worn once and thrown away, and a million or two was merely "loose change". The days of keeping a close eye on the stock market, of plotting and scheming to become the first "Trillionaire", will soon become a "dim and distant memory"!

As no one wants to lower their standard of living, it should come as no surprise that the billionaires are no exception. They regard Socialism, especially the Dictatorship of the Proletariat, as "hell on earth". After the revolution, we can expect them to go to any length to restore their "paradise lost". They will stoop to any depth, any level of deceit and deception. They will prey upon the weaknesses of the most honest. As they are superb liars, with life long experience, no doubt some working people will believe them. After all, there are so many people who believe that "there is a little good in everyone."

That may or may not be true, and it may or may not be a fine philosophy. It matters not, as the class struggle calls for Revolutionaries, not philosophers. We need tried and tested Revolutionaries, hardened veterans of the class conflict, those without any illusions, to be entrusted with the task of crushing the capitalists, as well as their most loyal and devoted followers.

For that reason, it will perhaps be best to isolate the billionaires, in remote areas, depriving them of any means of communication. At the same time, we want everyone to be useful, so perhaps a job in underground mining is something they can handle.

There are other members of that small but significant minority, and they are referred to as "class traitors". These people are members of the working class, yet at the same time, they are completely loyal to the billionaires. Under capitalism, for reasons which defy all rational explanation, these workers consistently lick the boots of the capitalists. They go out of their way to curry favour with the bosses, "ratting out" their fellow workers at every opportunity. The reward for such self degradation is absolutely nothing!

Under socialism, we can expect these same boot lickers to *defend* the billionaires, to demand that we return to them all the wealth they have stolen! In that case, perhaps it would be best to allow these class traitors to accompany their Lords and Masters. They can join them in the mines!

As for those who remain skeptical, doubtful that anyone could possibly be that stupid, may I remind you that several servants accompanied Czar Nicholas and his family, after he was removed from the throne and placed under arrest. Those same idiots also died with Nicholas and his family. I mention this as an example of extreme stupidity, in conjunction with a spiritual force.

This is to drive home the point that there are various "spiritual forces of oppression", to use the expression of Marx, or "invisible chains", which is my expression. These are in addition to the "physical force elements of repression", all part of the "existing state apparatus", once again according to Marx, all of which must be destroyed!

To destroy the physical is not terribly difficult. Brute force is required, and there is no shortage of working people who are experts in that department! Yet spiritual forces must also be destroyed, and brute force alone is not sufficient. The people who are in possession of these "spiritual forces of repression" must be crushed!

These spiritual forces are not to be underestimated! Just because they are invisible, does not mean that they do not exist! They merely supplement the physical force element, and if anything, are more dangerous! Because they are invisible, they are well hidden!

We must not repeat the mistakes of previous Revolutionaries, but learn from them. At the time of the Chinese Revolution, the spiritual power of the landlords, over the peasants, was broken. Yet the spiritual power of the University Professors, over all the common people, both peasants and workers, was not broken. This served to enable the Chinese capitalists to return to power, after the death of Mao.

At the time of the Second American Revolution, we can expect the spiritual power of the mobsters to be broken, just as the spiritual power of the Tories was broken, by their ancestors, at the time of the First American Revolution. Yet the spiritual power, the Professor Power, of the scientists and professional people, will also have to be broken. Otherwise, we run the risk of a return to power, by the capitalists.

For now, may the revolutionary battle cry be:

Prepare For Council Power and the Dictatorship of the Proletariat!

PART 8

RAISING THE LEVEL OF AWARENESS OF THE WORKING CLASS

Background

Over the course of the last few years, I have managed to identify a number of huge animals, which the scientists claim to be extinct.

This opens up various possibilities, opportunities for the people who are taking part in the revolutionary motion, to become ever more active. In fact, they can take part in several major scientific breakthroughs. The political struggle should not, and must not, be confined to demanding better wages and working conditions.

Before I proceed, I should mention that, for the purposes of this article, I use the word "capitalist" to refer to the super rich, the billionaires, technically referred to as the "bourgeoisie". It is important to distinguish them from the small time capitalist, the small business owner, technically referred to as the "petty bourgeois".

The working class has no quarrel with them, just as we have no quarrel with the professional, salaried employees. In fact, they are the natural and desirable allies of the proletariat.

I should also mention that I make a point of writing in a very popular manner, with working people in mind. This involves the explanation of scientific terms. Those who are already class conscious may find this to be tiresome, but it cannot be helped.

The members of the working class generally refer to themselves as "common people", although they have no objection to being referred to as "members of the public", or "working people", or the "little guy", or part of the "rank and file".

In fact, working people are technically referred to as "proletarians", although most of them are not aware of this. So now my goal is to raise their level of awareness, to make them aware of themselves as a class, with their own class interests. Those class interests are diametrically opposed to the interests of the monopoly capitalists, the billionaires, technically referred to as the "bourgeoisie". That which is in the best interest of the working class, is in the worst interest of the capitalists.

The best way to raise the level of awareness, of the working class, is by combining theory with experience. In terms of theory, I am encouraging all working people to read State and Revolution, by Lenin. That book explains the *necessity* of overthrowing the capitalists, of *smashing* the existing state apparatus, and *crushing* the capitalists, under the *Dictatorship of the Proletariat!*

Distinguishing Communists From Social Chauvinists

As a result of the revolutionary motion, countless working people, those who were formerly apathetic, are now becoming politically active. It is only to be expected that they should examine the various political parties, those which claim to be Socialist. It is also to be expected, that they should become confused!

The current crisis in capitalism has led to some interesting developments, which are to be expected, as they have happened so often in the past. The Revolutionary motion has had the unintentional consequence of highlighting two separate tendencies in International Marxism, otherwise known as Communism, formerly known as Social Democracy.

On the one hand, we have the true Communists, those who are determined to *overthrow* the ruling class of monopoly capitalists, the billionaires, to *smash* the existing state apparatus, and to set up a new state apparatus, in the form of the Dictatorship of the Proletariat.

On the other hand, there are the social chauvinists, those who claim to be Marxists, but who also maintain that Marxism must be "revised". Such people are of course referred to as "revisionists".

We refer to the social chauvinists as, "Socialists in words, chauvinists in deeds". They are determined that Communism must change from a Party of Social Revolution, into a democratic party of social reform. They reject the idea of *Scientific Socialism*, especially the Marxist theory of the *Dictatorship of the Proletariat!* The *touchstone* of a true Marxist is absolutely rejected! They think that Socialism and liberalism are the same thing! The *theory of the class struggle is rejected*, on the grounds that it cannot be applied to a democratic society, which implies "majority rule", and other such nonsense. Yet all too many of them insist on referring to themselves as Marxists!

Lenin dealt with such people extensively, in his book What Is To Be Done? He referred to them as Mensheviks, or Economists. They consistently choose the path of conciliation, instead of the path of struggle. He refers to such conciliators as those who are "in the swamp". They persist to this day, and in fact, there is no shortage of them! They insist that there is no need for a revolutionary theory, despite the fact that Engels stated, most emphatically, that *"without a revolutionary theory, there can be no revolutionary motion"*. In fact, Engels recognized three great struggles of Communism, *political, economic and theoretical!*

As working people, or at least the most advanced workers, have to be raised to the level of Marxists, I highly recommend a careful reading of that book.

With that in mind, perhaps it would be helpful to first give a brief historical background, as well as a translation of the Russian names.

At the time Lenin was writing this book, Marxism was referred to as "Social Democracy", as Marxists fight for democracy as well as socialism. Lenin was the leader of the Social Democratic Party. Yet the Party was split, into a majority, or Bolshevik, led by Lenin, and a minority, or Menshevik.

The Mensheviks were social chauvinists, revisionists, determined to revise Marxism, also "opportunists". This is to say that they were completely devoid of principle! It may help to read the word "opportunist" as "unprincipled". They were afraid to antagonize the capitalists! They wanted to reason with them! To appeal to their sense of humanity! To fight only for paltry reforms! Better wages, living and working conditions! Nothing else!

Lenin referred to this nonsense, this "class harmony", as "economism". It was with these hypocrites in mind, that Lenin wrote What Is To Be Done?

As for the Russian names, perhaps it would be helpful to read "Iskra", the newspaper for which Lenin wrote, as Spark. The newspaper "Robocheye Dyelo", means Workers Cause. "Rabochoye Mysl" translates as Workers Thought. "Zarya" means Sunrise. Then there is a reference to "Narodism", which is a form of agrarian socialism.

A careful reading of What Is To Be Done?, and other works by Lenin, such as State and Revolution, should provide a proper theoretical revolutionary basis. In this manner, working people will be able to determine the proper Marxists, Communists, from the social chauvinists, those who are determined to "patch up" capitalism.

As well, it is important to distinguish the Marxists, from those who claim to be Socialists, but not Marxists. Such self declared Socialists tend to fight for reforms, and are the allies of the Marxists, the true Communists. We certainly have no quarrel with such people! They tend to be people of principle, not hypocrites!

Looking For Lake Monsters

In the interest of gaining experience in the class warfare, I am suggesting that working people take part in challenging some "sacred scientific theories". This is not as difficult as it may appear!

Most common people are fascinated by various myths and legends, such as that of Sasquatch, Fire Breathing Dragons, UFO's, Flying Saucers and Ogopogo, among others. I maintain that these "legends" are based upon very real animals! I am further suggesting that, with a little effort, common people can take part in proving that they exist! This will have the effect of boosting their self esteem!

They will feel empowered! It will also serve to break the spiritual bonds of the scientists, the "professor power". So many working class people have placed scientists "on a pedestal"! To prove that the scientists are mistaken, will have the effect of empowering the workers! This will serve to break the spiritual bonds of the scientists! That which I refer to as "Professor Power"! As well, their self confidence will increase dramatically! It will inspire them to challenge other figures of authority, including the billionaires.

The precise method which is employed, in this noble endeavour, is critical. I can only suggest that the newly created Councils take the lead, in proving the existence of these animals.

Perhaps it would be best to start with proving the existence of the various "Lake Monsters", such as "Ogopogo".

In Okanagan Lake, this animal is referred to as "Ogopogo". I maintain that there are two of them. I also maintain that they follow the rivers, and are common place, located in a great many large lakes, in North America.

In Lake Champlain, they are known to locals as Champie. In Lake Erie, they are known as Bessie. Very likely each of the Great Lakes is home to these animals.

For that matter, it is very likely that the Loch Ness monster is none other than an ichthyosaur, or perhaps basilosaurus, if not both. There have also been reports of this animal located in other large lakes of Europe. Councils in those European countries are also advised to look for them. This I recommend!

May I suggest breaking people up into teams, perhaps with a mixture of young and old. Impress upon all that there is an element of danger in this, so that reasonable precautions are expected. Then each team should be assigned a sector of the lake, along with a map of the lake, and provided with a drone and numerous trail cameras.

The idea is that each team is responsible for sending their drone over the edge of the meadow, at the point where it meets the water. Mark out any "slide", which is the point where the animal enters and exits the water. Such slides should be quite obvious, devoid of any vegetation. Bare ground. Also look for any tree close to those slides. Be sure to mark the location on a map.

Such a task could well take all day. The following day, each team can go to the slides which they have located, and attach a trail camera to a tree, on the edge of the meadow. At least one member of the team should carry a high powered rifle. The idea is to protect the team, in case of an attack by a predator.

It is very likely that these trail cameras will be able to detect these animals, possibly when it enters the meadow, before full darkness, or in the moon light. After all, they are huge. Yet this also means that the cameras will have to be checked, and on a regular basis.

Of course, it is important to prove the existence of these animals. Yet the more important aspect of this little exercise, is to train people to work together, as a team, independently. At the same time, their performance can be quietly evaluated. This is valuable training for the revolution, and the subsequent Dictatorship of the Proletariat.

Necessity of Revolution

As for those who are skeptical, thinking perhaps that a revolution could not possibly happen here, feel free to face the facts. Bear in mind that the interests of the two classes, the workers and the billionaires, are diametrically opposed. The wealth of the capitalists comes at the expense of the workers. The more wealthy the capitalists become, the more impoverished the workers become.

Bear in mind that, according to the bourgeois economists, since the start of the pandemic, less than two years ago, the wealth of the billionaires has doubled! At the expense of the workers!

Those same bourgeois economists also report that most billionaires pay little or no taxes. Among other "tricks of the trade", they may "borrow" money, from their own businesses, as a means of supporting their lavish life style. This is not reported as income, so it is tax free! There are numerous other ways to avoid paying taxes, all perfectly legal!

Now there are a number of billionaires who are thought to be worth tens and even hundreds of billions! It is not enough! Each and every one of them wants to be a Trillionaire! In other words, they want to have the value of a thousand billion!

As one of these multi billionaires complained, in response to the suggestion that they should pay their "fair share" of taxes: "Eventually, they run out of other peoples money and then they come for you". The billionaires see themselves as victims!

Another billionaire stated it more prosaically, in response to the suggestion that he pay taxes, with a possible attempt to sound profound: "It is better to get humanity to Mars and preserve the light of consciousness".

Rather than pay taxes, in order to provide housing for the homeless, medical care for those who so desperately need it, food for the hungry, repairs to the roads and bridges, among a great many other things, he considers it more important to "preserve the light of consciousness" by sending "humanity to Mars"! Which is simply not possible! Yet the social chauvinists, Economists one and all, would have us believe that such people are prepared to submit to the will of the majority! Not likely!

The point is that the situation is truly revolutionary, as there is a limit to that which people are about to tolerate. With that in mind, may I suggest that some of these newly created Councils, especially those that are located close to huge fresh water lakes, take steps to verify the existence of these huge swimming animals. It is not difficult, and the results will be immediate and dramatic.

Councils, also known as Soviets

We can now consider the "Councils", a creation of the revolutionary movement. These are composed of leaders of the working class. Such people may be current or former members of the middle class. It matters not. The important thing is that they plot a course of action, for the working class.

People may not be aware of the fact that Soviet is a Russian word, which simply means Council. These Councils, or Soviets, are spontaneous creations of the working class, and first appeared in Russia in 1905, at the time of the First Russian Revolution. As they eventually gave birth to the Soviet Union, they are not to be under estimated!

It is not a coincidence that these Councils have also taken shape here, at this time. We too, are also on the eve of a revolution. We have got to be prepared, and there is no time to waste.

The government sees these Councils as a threat to their authority, as indeed they are. For that reason, they have to be discrete, or "work underground", as is the popular term.

In the city of Seattle, the Councils actually set up a Zone, and declared it to be Autonomous. They very quickly learned, just as the workers who took part in the Occupy Movement learned, that such Zones are not allowed! The Seattle Autonomous Zone was quickly crushed.

Yet the Councils remain! Now they are more discrete, working quietly, "behind the scenes"- properly so!- assisting the working people in various neighbourhoods. As well, working people are being armed and equipped, trained in the use of various weapons. In short, they are preparing for the next American Revolution!

In all cases, the Councils should mobilize as many working people as possible. It is to be expected that sports people, such as members of rod and gun clubs, will be especially interested in proving the existence of these huge animals. Then again, no doubt students will be equally interested, especially those who are students of science.

Yet as so many working people are interested in these legends, the response should be most impressive. There are other animals which have yet to be proven to exist. One of them is the Giants, or Gigantopithecus, otherwise known as Sasquatch or Bigfoot. It too, is quite easy to locate. They are on the Reserves along the Pacific Coast, and require the cooperation of the Indigenous people, but once that cooperation is given, then working people can prove the existence of another huge species. This will also provide the Councils with more valuable information, concerning the commitment and determination of workers.

Then there are the pterosaurs, otherwise known pterodactyls. They are commonly referred to as Dragons or Thunder Birds, although the list of local names is numerous, and locating them is more of a challenge.

May I suggest that people pay attention to any reports of cattle or horses that are found dead in the morning, terribly mutilated, but with no visible sign of blood. Then it is a matter of drawing out a sample of blood, as well as swabbing around the wound sites.

Both should then be rushed to a lab for analysis. The lab will then determine the poison gas that was used to kill the animal. Further, they will determine that the DNA is that of a reptile, one not known to science. Then it is a matter of getting permits to open up the old Ten Thousand Road. The entrance to the caves, which are the nesting ground of the pterodactyls, open up onto that Road. Then it is a simple matter of placing cameras at the entrance to the caves.

These are very simple tasks, which could well be carried out by the scientists. They are not carrying them out, because the last thing they want to do is to "disturb the peace and tranquility of the capitalists". That is the first thing we want to do! At the same time, workers will receive valuable training, and members of the Council will get a chance to determine the suitability of those workers, to be put to work, at the time of the Insurrection.

In particular, proof of the existence of walking whales and ichthyosaurs, will cause a major uproar. There will be an immediate call for the lakes, especially the Great Lakes, to be cleaned up. As they are severely polluted, mainly due to industrial run off, people will demand that the factories clean up their own mess. That is the last thing the capitalists want to hear! Perhaps if the scientists would spend less time squawking about climate change, and more time performing their duty, then we would live in a far better world!

Necessity of Insurrection

At the time of the revolution, first comes the insurrection, the day of reckoning. On that day, various key locations, across the country, will have to be secured by the revolutionary proletariat. It is not enough to take possession of the Vipers Nest, in the capitol of Washington, DC.

Mind you, that little task will probably not be terribly difficult, as the events of January 6 have revealed. But then the capitalists have also noticed the weakness in their defences. So they responded by building a wire fence around the buildings. Their childish faith in fences is somewhat touching, even if it is quite pathetic.

As I have documented in a previous article, on the day of the insurrection, it will also be necessary to shut down the railroads, bridges, tunnels, airports and sea ports, as well as communication networks. This is to say that all across the country, numerous groups of revolutionaries, working people, will have to take action, under the direction of a local Council. Each local Council must work under the supervision of a national authority. We will go into that detail later, in this article. Each and every one of these groups must be resolute. If even one group fails to carry out its assignment, the fate of the Insurrection could be in jeopardy.

That is where the quest for these huge animals can serve as a training exercise. May I suggest that as many Councils as possible, assign as many groups of workers as possible, in an attempt to locate these animals. The main thing is to determine the workers who are most determined, absolutely resolute. At the same time, take note of those who are somewhat indifferent. This is not to say that we are trying to judge people. It is to say that, on the day of the Insurrection, we must use only those who are completely resolute. We have no use, on that day, for those who are likely to waver.

Strikes and Revolution

It is entirely possible that the revolution has already started. Recently, a major airline cancelled hundreds of flights, over a period of several days. The press reports are rather vague, but there are references to people who refused to work, including pilots and air traffic controllers. Assuming that such people are members of a union, then such "walkouts" are referred to as "wildcat strikes". This is a reference to a strike which has not been authorized by the union leaders. If that is the case, then it is an indication of the strength of the revolutionary motion.

Bear in mind that most revolutions start with strikes in the transportation sector. That includes airlines, railroads and shipping lines. As the capitalists are complaining about delays in shipping, it is entirely possible that the workers are engaging in "slowdowns". That is not exactly a strike, but very likely a prelude to a strike. Such slowdowns take place as a result of deep worker dissatisfaction. As most union leaders are "in the pocket" of the capitalists, such dissatisfaction is completely understandable.

Strikes are one thing and insurrection is something else entirely! At some point, the working people who are taking part in the revolution have to seize political power! That calls for an insurrection!

This is not something to be taken lightly! The vast majority of the workers, or at least of the most advanced workers, must be prepared to overthrow the capitalists, smash the existing state apparatus, and establish the Dictatorship of the Proletariat!

The key word here is "prepare"! The most advanced workers must become class conscious, aware of the existence of classes. They must also become aware of the revolutionary theories of Marx and Lenin.

The necessity of smashing the existing state machine, and replacing it with the Dictatorship of the Proletariat, must be stressed.

All workers should be encouraged to read State and Revolution, by Lenin. A fine understanding of that superb book, will go a long way towards raising the level of awareness of the working class!

Another part of that preparation involves becoming organized, working together as a team, as part of a large army. The experience that working people are about to gain, in proving the existence of these huge animals, will prove to be valuable training.

Comparison to the Russian Great October Soviet Socialist Revolution

Bear in mind that in a similar situation, that of Russia of 1917, Lenin returned from exile in April. This was immediately after the Czar had been overthrown, and a democratic republic had been established. Yet Lenin did not immediately call for an insurrection.

In fact, a possible uprising in July of that year was aborted, as he thought that the working class was not properly prepared. Those days have gone down in history as the "Revolutionary July Days".

I mention this because it is so important. Lenin called off a possible insurrection, at that time, as it was clear to him that the working class, or at least the most advanced strata of the proletariat, had not yet embraced the Dictatorship of the Proletariat. This is another way of saying that the Russian proletariat was, at that time, not sufficiently class conscious. It was up to the Communists to raise the level of awareness, of the advanced workers, to that of Communists.

This they managed, which made possible the insurrection several months later, on October 25, old style calendar, or November 7, new style calendar.

The American proletariat of today is even less class conscious than the Russian proletariat of 1917, through no fault of their own. The conditions of life, of the proletariat, do not lead to the awareness of itself, as a class. This awareness must be brought to it, from an outside source. That is the duty of middle class intellectuals.

We clearly have our work cut out for us, but that is no cause for despair. Most working people are literate, and at least have access to digital devices. The task of raising their level of consciousness is far easier for us, than it was for the Russian Communists of 1917. Among other things, we have the internet, and we would be fools not to use it.

Working people should be encouraged to read State and Revolution, by Lenin. That is even available in audio form, and can be downloaded from the internet. As well, Left Wing Communism, An Infantile Disorder, is another most valuable book. It too, is available in audio form, and can be downloaded from the internet.

Those who are conscious people, aware of the revolutionary theories of Marx and Lenin, must flood social media with calls to read those revolutionary works, as well as with calls to become politically active! Combine theory with experience!

Bear in mind that, during times of revolutionary motion, countless workers, who were formerly apathetic, become politically active. As a result, the distinction between the most advanced and the less advanced, becomes less clear.

Yet the much less advanced workers must not be neglected. Popular literature must be made available to those workers, but by no means vulgar. Feel free to use sports metaphors, and avoid the use of the word "backward" when referring to workers. So many workers may consider this to mean "stupid", and the last thing we want to do is to offend any members of the working class. We want to flood social media with such literature. No doubt, among the less advanced, leaders will emerge.

We would all do well to remember that working class people are avid readers. They pay strict attention to the news, so that in the literature, be sure to use current examples.

I mention the Russian Great October Soviet Socialist Revolution, as that is the Revolution which most closely resembles our own.

Granted, there are considerable differences. The existence of the nobility, landlords and peasants, each with their own class interests, complicated the situation. Our revolution is simpler, in that we have the billionaires and the proletariat. All other classes have been all but wiped out.

This is to say that we live in a highly industrialized, or "cultured" country, as opposed to an under developed, or "petty bourgeois" country, commonly referred to as a "third world" country. I mention this for the sake of those who are just now becoming politically active. It also means that starting a revolution in this country is much more difficult, as the bourgeois ideology is so deeply entrenched. Yet carrying the revolution through, after the insurrection, is much easier.

Need For A Communist Party, Dictatorship of the Proletariat

It is also a fact that there is an urgent need for a true Communist Party, one which calls for the Dictatorship of the Proletariat. After all, people need leaders. Workers can only do so much! It is very likely that many members of the newly created Councils are conscious people, well educated, either current or former members of the middle class. Such people tend to be well aware of the revolutionary theories of Marx and Lenin. Precisely the sort of people we need to create a true Communist Party!

To such people, may I suggest that the creation of Councils is excellent, a step in the right direction, but merely a step. Half measures get us nowhere! The next step involves getting together with conscious people, Marxists, mainly from other Councils, and creating a true Communist Party.

No doubt, all members of the Councils consider themselves to be Socialists, or at least are sympathetic to Socialism. Equally without doubt, many of them are well aware of the revolutionary theories of Marx and Lenin. May I suggest that now is the time to apply those theories to a revolutionary situation. This is to say that a true Communist Party, one which calls for the Dictatorship of the Proletariat, is urgently needed.

The creation of such a Party may not be terribly difficult. No doubt, the various Councils are in touch with each other, so that the Communists on each Council can get together, not necessarily in person, and create a Party.

Avoid the use of the phone, as all such conversations are monitored. Even while using the internet, avoid certain words, as the government computers are programmed to "flag" such conversations.

Do not make anything easy for the government agents! Bear in mind that as the pedophiles are able to use the internet, while escaping detection, then so can we!

This is not to say that the Communist Party should take the place of the Councils. On the contrary, the Party should work as closely as possible with the Councils, as well as with the trade unions, cooperative societies and sports clubs. That same Party can also be the national authority, and at the time of the insurrection, coordinate the activities of all the local Councils.

There is no time to lose! Either the revolution will be led by Communists, or it will be led by reactionaries, such as Trump! Your choice! For the moment, the slogan from all conscious people must be:

Prepare For Council Power and the Dictatorship of the Proletariat!

PART 9

SCIENTISTS: FORM AN ASSOCIATION

As I write this, the trial, by the Senate, of President Trump is just starting. That is certainly a step in the right direction, and long over due. The importance of the trial lies not so much in the outcome, but in the fact that so many working people, including those who were formerly apathetic, are now taking a keen interest in the "democratic" process. They are becoming politically active.

It is also a fact that the ruling class, the billionaires, have arrived at a stage of crisis. They can no longer rule in the old way, and must change their method of rule. As yet, they are floundering, unable to decide precisely the new method of rule. The organization they have established, which I refer to as the White House Resistance, has yet to make a decisive move. This is to say that Trump has yet to be stopped. They are afraid that any move they make on Trump could trigger a full scale revolution. They are also afraid that if they do not stop Trump, he could trigger a full scale revolution!

Either way, revolution is on the horizon, and could break out at any day. Now is the time to raise the level of awareness of the revolutionary proletariat, to bring to the proletariat the awareness of itself, as a class.

The scientists are in a unique position to do this, in that they can document the lies and distortions of the capitalists, as written in the scientific textbooks. Some of these lies are quite flagrant. This may help to drive home the point that the billionaires are a pack of liars and hypocrites. As I have documented some of those lies in earlier articles, there is no need to repeat it here.

After the revolution, those lies will be exposed. Now is the time to get ahead of this, to join the revolutionary forces, and not be crushed under the wave of revolution, which is about to sweep North America.

No doubt all scientists are aware that to go public with these facts, would entail almost certain career suicide. This is certainly a legitimate concern, as most people have responsibilities, families who are counting on them for support. To such people I can only respond that there is strength in numbers. All members of the trade unions can testify to that!

Bear in mind that these unions were forced upon the working class. The capitalists gave the workers no choice! In much the same way, the capitalists are also giving the members of the middle class, the petty bourgeois, including the scientists, no choice other than to band together in a fellowship of some sort.

With that in mind, I can only suggest that scientists follow in the footsteps of the proletariat, and form an International Scientific Association. Such an Association would have real power, able to stand up to the billionaires, to demand legitimate change. Members of all branches of science should be encouraged to join the Association.

Possibly it could be extended to scholars and intellectuals. As the Association takes shape, the members can determine such little details. The broader the Association, the more power it will hold.

It should not be limited to one country, but to all countries of the world. Now that we have the internet, we would be fools not to take advantage of it. The capitalists are able to ruin the careers of any particular person, but not a whole Association of people!

Such an Association, or at least members of such an Association, could point out that Marx and Lenin were political scientists. They placed the class struggle on a scientific basis. It is doubtful, and probably not desirable, that the Association would advocate the Dictatorship of the Proletariat. After all, we want the Association to be as broad based as possible.

The Association would protect individual members, so that certain individuals could become politically active, calling for the Dictatorship of the Proletariat. As scientists tend to be highly respected, this endorsement would carry a great deal of weight. This would certainly help to raise the level of awareness of the working class. As a bonus, such scientists would not be a target of the revolution.

This measure of protection, of an Association, can be used to the advantage of those who are true Communists. Formerly, such dedicated "cadre" were encouraged to sacrifice their careers, and get a job as a proletarian. In this way, they were able to "integrate with the working people", and to "bring to the working class the revolutionary theories of Marx and Lenin".

May I suggest that it is no longer necessary to make such a sacrifice! Nor is it desirable! Now that we have the internet, there is no need to "fall on your sword"! Middle class Communists may prove to be far more useful in the positions they now hold! There is no need to get a job, as an hourly employee. The working class is well cultured! Almost all workers have computers and digital devices of various sorts! Now we can reach them, through various web sites. They can be encouraged to read key works of Marx and Lenin. Even State and Revolution can be downloaded from the internet!

This also applies to other members of the middle class. Make your friends and co-workers aware of the fact that only five businesses and eight banks, are Too Big To Fail! The other side of this coin, is that thousands of banks, and tens of thousands of businesses, are Too Small To Succeed! They are about to fail! We are facing a repeat of the Great Depression, on steroids!

As countless businesses fail, the stock market will collapse! Millions of workers will be unemployed! Such people cannot pay taxes! The billionaires do not pay taxes! Without any tax money coming in, the country is facing ruin! Unless the billionaires are overthrown!

When faced with these facts, it is quite possible that a great many middle class people may become revolutionary! After all, it is just a matter of time, before the middle class is completely impoverished! By the billionaires!

Do not wait for the axe to fall! Get ahead of this! At the very least, get involved with Councils! Be of assistance to the working people. Take part in arming, equipping and training the members, in preparation for the revolution. Get in touch with other Councils, and prepare for the insurrection. Perhaps most important of all, take part in the creation of a true Communist Party, Dictatorship of the Proletariat.

Middle class people, intellectuals with professional careers, are in a far better position to form a proper Communist Party, than working people, those with hourly jobs. It is doubtful that an hourly employee could manage to organize such a Party, although the most advanced workers can be most helpful.

Perhaps that requires a little explanation. All scientists can be said to be members of the middle class, petty bourgeois, and are no doubt, class conscious. They are aware of the existence of classes. The conditions of life, of the middle class, lead to this awareness. As well, all scientists have University degrees. Those degrees come complete with an exposure to the revolutionary theories of Marx and Lenin.

By contrast, the rather limited education of working people, and the life style of the working class, does not lead to the awareness of classes. These are merely facts that I am stating.

It is also a fact that the working class sometimes gets into revolutionary motion, as is currently happening. I refer to this as an Act of God. This is not to say that the members of the working class are conscious of their actions, because they are not. This is to say that the working class makes history. For that reason, the current ruling class, the billionaires, are about to be overthrown and crushed by the working class, under the Dictatorship of the Proletariat. The working class is, as yet, not aware of this!

At the time of the outbreak of the revolution, by which I mean a full scale uprising, the working people will attack those whom they consider to be the enemy. This includes the billionaires, although they are few in number. It also includes the people who support the billionaires, and they are quite numerous.

All scientists, and in fact all members of the middle class, would be well advised to distance themselves from the billionaires. Make it clear to the working class, before the "outbreak of hostilities", the revolution, that you are on the side of the proletariat. The alternative will not be pleasant.

With that in mind, I can only stress to scientists and intellectuals the urgency of forming an Association, if for no other reason than that of a safety net. Then, those who choose to become politically active can become involved with Councils, create a true Communist Party, and help to raise the level of awareness of the proletariat.

Those who choose to not become politically active run the risk of becoming targets of the revolution.

Either way, the revolution will soon break out. The higher the level of awareness of the working class, the smoother will be the transition to socialism, in the form of the Dictatorship of the Proletariat. It is to the advantage of all scientists to make sure that this transition is as smooth as possible.

I will close this article with slogans which I hope will soon become house hold expressions:

Workers of the World, Unite!
Scientists of the World, Unite!
Scientific Socialism!
Dictatorship of the Proletariat!

PART 10

MONOPOLY CAPITALISM

It is remarkable to think that for over thirty years, I have been working on this, my "little science project", as I refer to it. During that time, I have been challenging various scientific theories. This includes, but is not limited to, the theories of the mass extinctions of numerous orders of animals. The simplest way to disprove those theories, is by proving the existence of those same animals. Which is not to say that simple is easy!

For the benefit of those without a scientific background, I should add that an "order of animals" is defined as "a taxonomic rank used in classifying organisms, comprised of families sharing a set of similar nature or character".

It may help to think of the pterosaurs as an "order" of flying reptiles, composed of numerous species, located in different parts of the world.

There are currently a great many people who speak nostalgically of the time, before the age of Newton and Darwin, as the "good old days". It was certainly a much "simpler" time! It was accepted, and in fact it was insisted, that the world was at the centre of the universe! The answers to all questions, were to be found in the bible. Anyone who challenged this belief, could be charged with heresy and burned at the stake. It was this threat of execution, that served to limit scientific advancement.

All of that changed with Newton, and his three laws of motion. We now know that the earth is anything but the centre of the universe!

Other scientific break throughs followed. In particular, Darwin did for biology, that which Newton did for physics.

Not everyone was terribly happy about this. To this day, there is no shortage of people who resist change! As well, there are a great many people who object to the theory of evolution, based on religious beliefs.

In particular, those who are referred to as "fundamentalists", or "creationists", deeply disapprove of the theory of evolution.

The beliefs of these people, common people, members of the public, must be respected. It is their religion. To even suggest that apes and humans have a common ancestor, is to insult them!

That being said, it is important to separate science and religion. Religious beliefs have no place in the class room! To teach "creation science" in the schools, as an "alternative" to the theory of evolution, cannot be allowed. After all, "creation" is a belief, not a scientific theory.

As that is the case, it is essential that we distinguish the fundamentalists, from the reactionaries. The fundamentalist have their religious beliefs, while the reactionaries are those who want "everything to stay exactly the way it is". Such people are opposed to any social or political reform.

As for those who are of the opinion that such people are rather quaint, harmless sorts, you are mistaken. Feel free to face the facts.

Our Civilization In Decline

Over the last several thousand years, numerous civilizations have come into existence, flourished, risen to a peak, and then fallen into decline. This decline in civilization is not an Act of God. It is an act of people! Honest, hard working people, build up a civilization.

As opposed to this, are the reactionaries. They are lying, thieving, destructive people, intent only on tearing down anything that has been created.

Under our present civilization, our ancestors have worked all their lives to build a better world. They made great sacrifices for us, their descendants. Some of them made the "ultimate sacrifice", while others worked themselves into an early grave. Without doubt, they built this civilization. Equally without doubt, it has now passed its peak, and is in decline.

I can only stress that this decline is an act of people, reactionaries, not an Act of God. Unless we take acton now, our civilization will go the way of all previous civilizations, the "way of the dodo bird".

The reactionaries would have us believe that we are "destined" to fall into decline, to join all previous civilizations. Such is hardly the case!

Our civilization is exceptional, in that we have experienced an industrial revolution! In the process, two new classes were created. The bourgeoisie, and the proletariat. Both classes have played a revolutionary role.

Marx and Engels have made that quite clear, in the Communist Manifesto:
"The bourgeoisie, historically, has played a most revolutionary part.

"The bourgeoisie, wherever it has got the upper hand, has put an end to all feudal, patriarchal, idyllic relations...It has resolved personal worth into exchange value...has stripped of its halo every occupation hitherto honoured.... has torn away from the family its sentimental veil, and has reduced the family relation to a mere money relation".

That is "on the one hand", so to speak. The Communist Manifesto goes on to say:

"The bourgeoisie...has been the first to show what man's activity can bring about. It has accomplished wonders, far surpassing the Egyptian pyramids, Roman aqueducts and Gothic cathedrals...The bourgeoisie cannot exist without constantly revolutionizing the instruments of production ...uninterrupted disturbance of all social conditions..."

The Manifesto draws a stark contrast to all "earlier epochs", before the industrial revolution, in which the "first condition of existence" was in maintaining the "old modes of production in unaltered form". This is to say that previous civilizations were determined that nothing should change.

This Manifesto was written in 1848, at a time when capitalism was still at the stage of competition. That changed, around the beginning of the twentieth century, at which point capitalism reached the stage of monopoly, referred to as imperialism. Not too surprising, monopoly capitalism, imperialism, has characteristics which are different from those of capitalism in its early, competitive stage.

It was Lenin who subjected imperialism to a careful analysis, in his book, Imperialism, the Highest Stage of Capitalism. He stressed the point that imperialism, is "reaction, right down the line". That is another book that deserves careful consideration.

Yet the industrial revolution gave birth to a second class. That other class is the class of proletarians, the working class. The proletariat has nothing to sell but their labour power. This class too, is revolutionary. This is the class that will turn around the decline in our civilization.

I can only stress the fact that all previous civilizations fell into decline, because there was no revolutionary class to reverse that decline. None of those civilizations passed through an industrial revolution, so that no proletarian class was created.

Decline In Science

An indication of the decline in our civilization, is to be clearly seen in science. Formerly, scientific theories were challenged, properly so. That is the proper scientific method! That is the one and only way to determine if a theory is correct!

That stands in stark contrast to the present day, in which all too many theories are considered to be "sacred", not to be challenged! That includes the "mass extinction" of so many orders of animals, which never happened!

This retrograde trend in science is nothing other than the result of the monopoly capitalists, the imperialists, all of whom are reactionaries, interfering in science.

Mind you, they do not refer to themselves as capitalists. They refer to themselves as "entrepreneurs", or "merchants", "business people". As if changing the name changes the nature of the beast! They live "in the past", and are determined that nothing will change!

The science books have been written, and are not to be rewritten! The same is true of the history books and all text books used in the schools! The fact that all of those books are filled with distortions and outright lies, does not impress them. The current glorification of slave owners, such as Jefferson and Washington, is merely an indication of the mind set of reactionaries.

As part of their method of rule, the capitalists have established very high rates of tuition, for university training. This has given rise to very expensive student loans. Further, any and all students who wish to pursue a career in science, must not "rock the boat". This is to say that they must not challenge the theories that are presented.

The penalty for such "heresy" is "career suicide". In other words, they are not allowed to earn a living working in their chosen field. This is referred to as being "black balled". At the same time, they must still find a way to earn a living and repay the loans.

In this way, scientists are degraded. They are forced to embrace theories that they know to be false, as that is the only way to survive.

By way of exception, there are the few students who are members of families who have enough wealth, to pay for the school tuition. Such students are supremely class conscious. They know better than to "rock the boat"!

Most children of very wealthy families, usually do not pursue a career in science. Such families are referred to as "upper middle class", or "upper class", bourgeoisie. The students are generally focused on careers in business, law and politics. Such careers tend to overlap.

For that reason, most of those whom earn a degree in science, are of "middle class" background, which is to say, petty bourgeois. They also generally graduate deep in debt! This helps to explain the silence of the scientists! The only way they can pay off the huge student debt, is by working in their chosen field of science! The only way they can work in their chosen field, is by not "rocking the boat"!

As a result of this, most common people, members of the public, are blissfully unaware of the existence of these prehistoric animals, including the pterosaurs.

It is the pterosaurs, in particular, that frequently prey upon people. They hunt in open areas, in the darkness, and as predators, have a keen sense of smell. It is characteristic of predators that they are far more likely to attack when they smell blood. That is the reason it mainly preys upon girls of childbearing age. It also preys upon children, as they are easier to pick up and carry away.

No doubt, the scientists are aware of the existence of these huge species. Equally without doubt, they remain silent in order to protect their careers.

I have no career to protect, as I committed career suicide many years ago. It was the price I paid for "rocking the boat". For that reason, the capitalists cannot threaten me with career suicide. Besides, my days of "rocking the boat" are over. Now I am "blowing the boat right out of the water"!

Soon, a great many careers will be ruined, or at least severely compromised, as we prove the existence of these huge animals.

As the scientists either cannot, or will not, challenge any scientific theories, it is now up to the common people to "blaze the trail". To prove the existence of these species is not difficult.

Attack On Our Democratic Rights

Then again, the revolutionary motion which is sweeping the world, is growing ever stronger. It is just a matter of time before the scientists are affected by this revolution. When that happens, we can expect them to rise up, band together and demand change.

No doubt, many readers are of the opinion that I am overstating the case. So for the benefit of such simple, optimistic, misguided souls, may I refer you to an American film maker, Michael Moore.

Mr. Moore is a most patriotic American. He has produced a number of documentaries, of which I consider, Capitalism: A Love Story, to be his masterpiece. It is important to note that Mr. Moore makes no claim to be a Marxist.

In his film, he refers to the monopoly capitalists, the billionaires, the bourgeoisie, as the "one percent". That is an expression the common people came up with, at the time of the Occupy Movement. They also referred to themselves as the "ninety nine percent". It was a step towards class consciousness, but only a step. They were instinctively "drawing a line" between themselves, the working class, and the "super rich", the billionaires, the bourgeoisie.

Having said that, I have attempted to summarize the facts presented on the documentary. Bear in mind that Moore makes no reference to classes.

In his film, Moore said that he managed to get his hands on a secret Citibank memo, concerning the plan of the "one percent" to rule the world. He maintains that Citibank wrote no less than three memos to their wealthiest investors. I have done my best to summarize that which was stated in the documentary:

"The bank has concluded that the United States is no longer a democracy, but a plutocracy a society controlled exclusively by and for the top one percent of the population. As they point out, the top one percent of the citizens have control of as much wealth as the bottom ninety five percent combined, and the gap is growing. Their biggest concern is that the ninety nine percent may demand a more equitable share of the wealth.

"The one percent are now the new aristocracy, but their big concern is that the ninety nine percent may revolt. They lament that the ninety nine percent still have the right to vote."

That was the conclusions that Moore was able to draw, from those memos. Moore also makes the point that the Constitution makes no reference to capitalism, or to the right of the capitalists to make a profit.

He could have further made the point that the Founding Fathers of the United States, those who wrote the Declaration of Independence, went much further. As they stated it, "We hold these truths to be self evident, that all men are created equal, that they are endowed with their Creator with certain inalienable rights...whenever any form of government becomes destructive of those ends, it is the right of the people to alter or abolish it, and to institute a new government."

The Declaration of Independence gives the American people the right to abolish any government which does not represent them! This is a point which must be stressed to all Americans! They are one of the few people, if not the only people, who have that right! But then they have a revolutionary history, of which they can be most proud!

Now it is a matter of building upon that revolutionary history.

No doubt the bourgeoisie would love nothing better than to scrap the Constitution and the Declaration of Independence. Those revolutionary documents stand for democracy, in the form of majority rule, and the capitalists are dead set opposed to such democracy.

As for those of us who have lived all our lives under capitalism, which is almost everyone, it is only natural to consider this as a normal state of affairs. Yet it is anything but the normal state of affairs. In fact, it is supremely abnormal.

Capitalism first came into existence, roughly three hundred years ago. It was the industrial revolution that gave birth to capitalism.

Before that time, there were no capitalists, bourgeoisie, just as there were no workers, proletarians. The point being that capitalism is a relatively new creation. Further, it was created by people. Capitalism is not an "Act of God"! It was created by people, and it will be destroyed by people! Capitalism will be destroyed by one of the classes of people it created, the working class, the proletariat.

Marx and Engels documented the dual nature of capitalism in their landmark work, The Communist Manifesto. They reveal to all, the progressive aspects of capitalism, at least in its early stages, as well as its reactionary features. This is referred to as the "dual nature" of capitalism. It is hoped that all readers will read that essential work of Scientific Socialism, the only true socialism.

Class Struggle

The Communist Manifesto was written in 1848. Of particular interest is the introduction to that work, by Engels, in 1883, shortly after the death of Marx. As Engles stated:

"The basic thought running through the Manifesto -that economic production, and the structure of society at every historical epoch necessarily arising therefrom, constitute the foundation for the political and intellectual history of that epochs; that consequently (ever since the dissolution of the primeval communal ownership of the land) all history has been a history of class struggles, of struggles between the exploited and the exploiting, between dominated and dominating classes at various stages of social evolution; that this struggle, however, has now reached a stage where the exploited and oppressed class (the proletariat) can no longer emancipate itself from the class which exploits and oppresses it (the bourgeoisie) without at the same time forever freeing the whole of society from exploitation, oppression, class struggles -this basic thought belongs solely and exclusively to Marx."

Shortly after that, capitalism reached the stage of monopoly, of imperialism, which is complete reaction. This is to say that we can expect the decay of our civilization.

To counteract that, we must prepare the working class, the proletariat, for revolution and the subsequent Dictatorship of the Proletariat. With that in mind, we should stress that our "economic production" is now socialized, so that our political structure must fall into line.

Scientific Socialism must be established. This is to say that it must be Marxist, as it has been proven that utopian socialism does not work. The monopoly capitalists, the bourgeoisie, the billionaires, are not about to surrender their hard stolen wealth and power, without a fight. That is the reason they must be overthrown and crushed, under the Dictatorship of the Proletariat.

I consider the quest for these huge species to be part of the revolutionary movement. We are entitled to our wildlife. It is part of our heritage.

It is frequently stated that any great scientific break through, will have to be made by youngsters. Those who hold that belief, usually use the example of Newton. They point to the fact that he did all of his scientific work before the age of twenty one. That is true, but it does not mean that the rest of us should retire our brains at an early age! I am certainly not a youngster, and am physically not capable of doing that which I did many years ago. But then my brain has not atrophied. I just hope that others, young and not so young, will be motivated to take part in a scientific break through. It is not something you will ever regret.

I can only stress that this is part of the revolutionary movement. Working people must become revolutionary. This is a fine place to start.

PART 11

THE INDUSTRIAL REVOLUTION

Recently, it has been brought to my attention that over the years, numerous civilizations have come into existence. They all prospered, became ever more powerful, rose to a peak, and then fell into decline.

In the western world, the Roman Empire is supremely well known. As the Roman Empire lasted for hundreds of years, people in those days were fond of saying that "Rome is Eternal".

Of course, nothing is "eternal", and the fate of the Roman Empire is well known. It went the "way of all empires", so to speak. Yet that does not mean that our civilization must "follow suit", as so many "philosophers" maintain. There is a huge difference between our civilization and all previous civilizations. Our civilization is the one and only civilization to have experienced an industrial revolution!

Historians are agreed that the industrial revolution was the greatest thing to happen to humanity, since the domestication of plants and animals. In this, they are absolutely correct!

With that in mind, perhaps a little explanation is in order.

The industrial revolution first took place in Great Britain between the years of 1720 to 1760, for reasons which do not directly concern us. From there, it spread to other parts of the world, and in fact, is still spreading.

At that time, in Britain, there existed a small, rather unimportant class of people, merchants who lived in town and referred to themselves as "burghers". They regarded the industrial revolution as an "opportunity", a chance to become supremely wealthy. It was merely a matter of investing their money in factories, mills, mines and other "means of production", as well as railroads, shipping lines and other "means of transportation". In that way, the "raw materials" can be taken to the "point of production", and the "finished product" can be "taken to market". They also invested in banks and other "financial institutions".

I believe those are the correct technical expressions of the capitalists.

As a result of this, the class of burghers became transformed into a class of capitalists. The name burgher became altered to that of "bourgeois". Those who became supremely rich became known as the "bourgeoisie".

Incidentally, some readers may find these technical terms to be tiresome, and perhaps they are. Yet it is important to become familiar with them, as otherwise, our class enemies will use our lack of knowledge against us.

For the purposes of this article, I am mainly concerned with the fact that the industrial revolution gave birth to two different classes, both revolutionary, at least at first!

Creation of New Revolutionary Classes

As Marx and Engels stated, quite clearly, in the Communist Manifesto: "The history of all hitherto existing society is the history of class struggles.

"Freeman and slave, patrician and plebeian, lord and serf, guild master and journeyman, in a word, oppressor and oppressed, stood in constant opposition to one another, carried on an uninterrupted, now hidden, now open fight, a fight that each time ended, either in a revolutionary reconstitution of society at large, or of the common ruin of the contending classes.

"In the earlier epochs of history, we find almost everywhere a complicated arrangement of society into various orders, a manifold gradation of social rank. In ancient Rome, we have patricians, knights, plebeians, slaves; in the Middle Ages, feudal lords, vassals, guild masters, journeymen, apprentices, serfs; in almost all of these classes, again, subordinate gradations".

It is clear that in the case of the Roman Empire, for example, the "fight" of the "contending classes" ended in the "common ruin of the contending classes", as there was no "revolutionary reconstitution of society at large". In short, the Roman Empire rose to a peak, fell into decline, and eventually rotted away. This is characteristic of most civilization.

It is also clear that our civilization has also passed its peak, and is now in decline. Our roads, bridges, railways and transportation network, as well as our buildings, that which is usually referred to as the "super structure", is in desperate need of repair.

Our books of science, which are taught in schools, are nothing less than a farce. The existence of classes, at least here in North America, is denied. Yet that does not mean that "we are doomed!"

The Communist Manifesto makes it quite clear that, as a result of the industrial revolution, radical changes were established in society:

"The modern bourgeois society that has sprung from the ruins of feudal society has not done away with class antagonisms. It has but established new classes, new conditions of oppression, new forms of struggle in place of the old ones.

"Our epoch, the epoch of the bourgeoisie, possesses however, this one distinctive feature: it has simplified the class antagonisms. Society as a whole is more and more splitting up into two great hostile camps, into two great classes directly facing each other: bourgeois and proletariat."

This is especially true now, as around the beginning of the twentieth century, capitalism reached the stage of monopoly, technically referred to as "imperialism".

The monopoly capitalists, the imperialists, are determined to crush any competition, no matter how insignificant. As a result of this, the small time capitalist, which is to say the middle class small business owner or "petty bourgeois", has largely been wiped out, at least in the most highly industrialized countries of the world. The same is true of the family farmers, otherwise known as "peasants".

That brings us to the "intellectuals and salaried personnel" of the capitalists. Lenin says that they "correspond to the middle class".

They tend to lead lives of "quiet desperation", waiting for the "axe to fall". Regardless of how well they do their job, they are well aware of the fact that the capitalist may fire them at any time, for any reason, or for no reason. "Because they can"!

Capitalists At First Revolutionary

The Communist Manifesto goes on to state:

"Modern industry has established the world market...and in proportion as industry, commerce, navigation, railways extended, in the same proportion the bourgeoisie developed, increased its capital, and pushed into the background every class handed down from the middle ages.

"We see, therefore, how the modern bourgeoisie is itself the product of a long course of development, of a series of revolutions in the modes of production and exchange.

"Each step in the development of the bourgeoisie was accompanied by a corresponding political advance of that class....

"The bourgeoisie, historically, has played a most revolutionary part.

"The bourgeoisie, wherever it has got the upper hand, has put an end to all feudal, patriarchal, idyllic relations. It has pitilessly torn asunder the motley feudal ties that bound man to his 'natural superiors', and has left remaining no other nexus between man and man, than naked self interest, than callous 'cash payment'. It has drowned the most heavenly ecstasies of religious fervour, of chivalrous enthusiasm, of philistine sentimentalism, in the icy waters of egotistical calculation. It has resolved personal worth into exchange value, and in place of the numberless indefeasible chartered freedoms, has set up that single, unconscionable freedom - free trade. In one word, for exploitation

"The bourgeoisie has stripped of its halo every occupation hitherto honoured and looked up to with reverent awe. It has converted the physician, the lawyer, the priest, the poet, the man of science, into its paid wage labourers.

"The bourgeoisie has torn away from the family its sentimental veil, and has reduced the family relation to a mere money relation…It has accomplished wonders far surpassing Egyptian pyramids, Roman aqueducts and Gothic cathedrals; it has conducted expeditions that put in the shade all former exoduses of nations and crusades

"The bourgeoisie cannot exist without constantly revolutionizing the instruments of production, and thereby the relations of production, and with them the whole relations of society...

"The bourgeoisie has through its exploitation of the world market given a cosmopolitan character to production and consumption in every country...

"The bourgeoisie, by the rapid improvement of all instruments of production, by the immensely facilitated means of communication, draws all, even the most barbarian, nations into civilization...

"The bourgeoisie.. has created more massive and more colossal productive forces than have all preceding generations together...

"We see then: the means of production and of exchange, on whose foundation the bourgeoisie built itself up, were generated in feudal society...

the feudal relations of property became no longer compatible with the already developed productive forces...they were burst asunder...

"Into their place stepped free competition, accompanied by a social and political constitution adapted to it, and by the economical and political sway of the bourgeois class."

I have chosen to quote the Communist Manifesto at length, partly because it is so important, and partly to drive home the point that at one time, the bourgeoisie played a most revolutionary role!

Monopoly Capitalism, Imperialism

Such is no longer the case! As soon as capitalism reached the stage of monopoly, which is technically referred to as "imperialism", the monopoly capitalists, the bourgeoisie, became completely counter revolutionary, referred to as "reactionary". Lenin documented this supremely well in his book, Imperialism, the Highest Stage of Capitalism.

These days, with the bourgeoisie so completely reactionary, we tend to lose sight of the fact that, at first, the capitalists were truly revolutionary! But to paraphrase an old and tired expression, "that was then and this is now"!

The Communist Manifesto goes into this in more detail:"Modern bourgeois society with its relations of production, of exchange and of property, a society which has conjured up such gigantic means of production and exchange, is like the sorcerer, who is no longer able to control the powers of the nether world whom he has called up by his spells. ...It is enough to mention the commercial crises that by their periodical return put on trial, each time more threateningly, the existence of the entire bourgeois society...an epidemic of over production..."

The industrial revolution made possible, for the first time in history, the chance to provide for the well being of countless people. Everyone has a chance to benefit from this vast surplus! But as long as the monopoly capitalists are in charge, that is not about to happen! It would never occur to them to provide for the common good! They are completely focused on their "bottom line", which is the very thing they call their sacred profit!

Yet under capitalism, a "vast surplus" gives rise to "an epidemic of over production", a "crisis in capitalism". It threatens their profits! As a result, even less is available for the common people!

That is the "one side of the coin", so to speak. Scientific Socialists may say that this is but "one aspect of the contradiction". The "flip side of the coin", or the "other aspect to the contradiction", is the fact which is stated quite clearly in the Communist Manifesto:

"The weapons with which the bourgeoisie felled feudalism to the ground are now turned against the bourgeoisie itself

"But not only has the bourgeoisie forged the weapons that bring death to itself; it has also called into existence the men who are to wield those weapons - the modern working class - the proletarians."

The fact of the matter is that as capitalism becomes ever more highly developed, so too does the modern working class, the proletariat.

At least here in North America, and probably in most other highly developed countries, the proletariat is quite highly cultured. The vast majority are literate, and either own computers, or some other "digital device", or at least have access to them. This makes it ever so much easier to raise their level of awareness. The capitalists have thoughtfully created the Internet. The least we can do is express our appreciation, by using it against them!

Women In the Work Force

Those who have been paying even the slightest bit of attention to the news lately, are no doubt struck by the fact that women are now playing a vital role in the international working class movement.

In particular, a young woman in Iran was arrested and killed, by the "morality police", for the crime of "not wearing her hijab properly". Bear in mind that a "hijab" is a head scarf.

The protests against such violent repression have spread across Iran, to all cities and towns, and even to remote areas. Videos from Iran show women dancing in the streets, waving their head scarfs, burning those same scarfs and cutting their hair. These protests have spread to other countries, so that women in various countries are supporting their sisters in Iran.

The Communist Manifesto helps to explain this also:

"The more modern industry becomes developed, the more is the labour of men superseded by that of women."

As ever more women become drawn into the work force, becoming proletarians, the more they become revolutionary. After all, it is the proletariat which is the most revolutionary class!

This infusion of "fresh blood" into the international working class movement is most welcome! No doubt many of these women have been "watching from the sidelines", so to speak, perhaps "marvelling at the stupidity of so many men". Or so I have been told!

Now those same women, those whom were formerly confined to the drudgery of housework, have "come into the work force", have become proletarians, and therefore revolutionary, and are making their voices heard.

The fact is that a great many working men are supremely well aware that most of our leaders, including our union leaders, are "in the pocket of the capitalists". Yet by and large, most men tolerate this.

By contrast, the women whom have recently entered the workforce, are not at all inclined to accept the "status quo". Now that they are workers, proletarians, they are becoming ever more politically active. They are a most welcome edition to the working class, a "breath of fresh air"!

Without doubt, here in North America, they are in the forefront of the working class movement.

Capitalism In Crisis

This brings us to the current political situation, in which capitalism is definitely in a state of crisis. The capitalists can no longer rule in the old way, and the working people are no longer content to be ruled in the old way. Lenin refers to this as a revolutionary situation.

Inflation is spiralling out of control, to the point that even the bourgeois economists are worried of a "repeat of the Weimar Republic", in which the value of the German mark depreciated, to the point that it was worthless. Those same economists are also openly speaking of a recession, while as yet, avoiding any mention of a depression.

Gun violence is so common, it is not even documented, unless four or more people are shot. Drug overdoses are now epidemic. The criminal gangs are in control of many neighbourhoods. Local police are reduced to documenting the crimes. A great many police officers are retiring or simply quitting. Very few people are even applying to join any police force, as "no one wants to become a cop". There are calls for "martial law" in some cities. It has been suggested that only the National Guard can "restore law and order".

From the viewpoint of the capitalists, the "cherry on the cake", is the fact that "one of their own", by whom they of course mean the former president, Donald Trump, is causing them untold grief! They just do not know what to do with that boy! He just will not shut up!

Recently, the New York Attorney General, NYAG, issued a two hundred page report, the product of a three year investigation. In that report, she alleges that Trump, along with three of his adult children, over a ten year period, submitted "fraudulent and misleading financial statements", on a regular basis. Accordingly, she is suing them for a minimum of 250 million dollars. Yet there are no criminal charges! The government is afraid of Trump!

In a very clear cut example of "passing the buck", she mentioned that the case had been referred to "federal prosecutors and the Internal Revenue Service", the IRS, for "possible federal crimes".

By implication, all of those allegedly "fraudulent and misleading financial statements" failed to break any state laws! Not likely! She just dares not charge Trump!

If any working class person was suspected of committing such acts, especially that of tax evasion, then that person would likely never again see the light of day! Without doubt, there is a double standard here!

In fact, a federal judge admitted as much. The opinion she gave is a masterpiece of legal jibberish: "Based on the nature of this action, the principles of equity require the Court to consider the specific context at issue, and that consideration is inherently impacted by the position formerly held by the Plaintiff".

Now to put this in plain and simple English, she refers to herself as the Court, and to Trump as the Plaintiff. Her reference to her "consideration" of the "specific context at issue" which is "inherently impacted by the position formerly held" by Trump, is a reference to the fact that he was formerly the president. For that reason, she favoured Trump! So much for the courts being fair and impartial!

Clearly, the laws do not apply to the capitalists, only to the working people. No kidding! Yet this federal judge broke the "unwritten law" of all courts: Do not admit this!

Most members of the working class are well aware of the fact that the laws do not apply to the wealthy. Those who have enough money are quite capable of buying themselves out of any criminal charges. There is no need to explain this to them!

By contrast, there is an urgent need to explain to the working people, the existence of classes. They must be advised, in terms they can understand, that those of us who work for wages are working class, "proletarians". The billionaires, monopoly capitalists, those who own almost everything of any great value, are members of a different class, that of the "bourgeoisie". Further, we are class enemies. The capitalists must be overthrown, the existing state apparatus destroyed, and the capitalists must be crushed, under the Dictatorship of the Proletariat. Yet who is to bring them this awareness?

The Communist Manifesto provides us with the answer: "Entire sections of the ruling classes are, by the advance of industry, precipitated into the proletariat, or at least threatened in their conditions of existence. These also supply the proletariat with fresh elements of enlightenment and progress.

"Finally, in times when the class struggle nears the decisive hour, the process of dissolution going on within the ruling class, in fact within the whole range of old society, assumes such a violent, glaring character, that a small section of the ruling class cuts itself adrift, and joins the revolutionary class, the class that holds the future in its hands. Just as, therefore, at an earlier period, a section of the nobility went over to the bourgeoisie, so now a portion of the bourgeoise goes over to the proletariat, and in particular, a portion of the bourgeois ideologists, who have raised themselves to the level of comprehending theoretically the historical movement as a whole.

"Of all the classes that stand face to face with the bourgeoisie today, the proletariat alone is a really revolutionary class. The other classes decay and finally disappear in the face of modern industry; the proletariat is its special and essential product."

As I write this, the capitalists do not know "which way to jump". They are currently "counting down the days" to the midterm elections. Perhaps they think that a new Congress will solve all their problems!

The Democrats are fretting over losing control of the House or the Senate, while at the same time worrying that Trump will once again run for president, in the next federal election, two years from now.

As for the revolutionary motion, they think perhaps the best way to divert it is with their "brainstorm" of "30x30". This is what Biden refers to as setting a goal of "conserving at least 30 percent of our lands and waters by 2030". Naturally, he was careful to not specify precisely what he meant by "conserve"!

This was pointed out in an excellent article of National Geographic. They provide an example of a farmer, a small time capitalist, who was quite happy to cooperate, to create "70 acres permanent wetlands", in the interests of conservation.

Of course, that came with a government grant of "$350,000!" At taxpayer expense, of course! As the farmer stated, "it has to pencil out". By that rather strange expression, he meant that he was not about to sacrifice profit for wildlife. As long as the taxpayers are prepared to hand over money to small time capitalists, they in turn are happy to cooperate!

So much for the Democrats plan for "conservation"!

The Republicans are also "scratching their heads", trying to come up with something to counteract Trump and his "Make American Great Again, or MAGA". It should be noted that the keystone of the MAGA policy is to hold immigrants responsible for the horrors of capitalism.

The fact is that even the journalists, those faithful flunkies of the capitalists, are openly mentioning the possibility of civil war. It does not help that Trump recently issued a "thinly veiled threat" to that effect. He made it quite clear that if he is charged, his followers are prepared to take up arms!

The capitalists are taking him seriously! They are afraid of Trump! Each government agency is "passing the buck"! The state of New York is going after him for hundreds of millions which they say he owes them, while documenting countless instances of fraud and tax evasion, over the last ten years, but not filing charges!

It is clear that the capitalists are trying to discredit Trump, to expose him as a liar and fraud. That is by no means a "tall order", as that is precisely the case! In other words, Trump is a typical capitalist! The more they expose Trump, the more they expose themselves!

The capitalists are trying to turn "public opinion", as they phrase it, against Trump! Excellent! By a remarkable coincidence, we too are attempting to do the same thing! In this way, we can raise the level of awareness of the working class!

The capitalists are not about to face the fact - or are not capable of facing the fact? - that we are in the middle of a revolutionary situation. Yet even the journalists, those most devoted servants of the capitalists, now dare to mention this!

True, capitalism has reached a state of crisis. That is a fact. It is also a fact that the system will not collapse upon itself! So for those who think that the capitalists cannot find a way out of this mess, think again!

As long as the proletariat does nothing, the capitalists will manage! It is a fundamental tenet of Marxism, that all reactionaries are the same! They must be destroyed! They are not about to simply "fade away!" The capitalists are reactionaries, right to the very core of their being, and must be crushed!

It is the working class, the proletariat, and only the proletariat, that is up to this task. After all, it is only the proletariat that is the consistently revolutionary class. Yet that is not enough. Their level of awareness has to be raised. They need to be made aware of the existence of classes, of the conflict between the classes. Above all, they must be made aware of the necessity of revolution and the subsequent Dictatorship of the Proletariat.

Contrary to what the demagogues may say, working people need leaders. Proper leaders! The only proper leaders are Scientific Socialists, Marxists, Communists!

The fact of the matter is that all classes have leaders, and that includes the capitalists. At the moment, one of the leaders of the capitalists is a fellow by the name of Donald Trump! He has quite a following, among the working people! But then, Trump is a first class demagogue!

Such people are not to be under estimated! Demagogues are the worst enemies of the working class! They have quite a following, among the proletariat, especially with the less advanced! The less advanced workers are not to be blamed for this! They are under the influence of those who peddle the lies of the capitalists! They have been listening to such lies all their lives! Who can blame them for believing those lies?

These working people need to be told the truth. Further, it must be expressed in terms which they can understand. This is to say that we must use popular language, not "High English"! Feel free to use sports metaphors, but under no circumstances resort to vulgarity!

Our goal it to raise their level of awareness, not to personally sink to a lower level. Also, avoid the use of the word "backward". They may think that we are calling them "stupid".

There is no time to lose. The revolutionary motion is becoming ever stronger. Civil war could break out at any day. Now is the time to become active. Take part in the emerging Councils. Join the mainstream political parties, as card carrying members. Run for office. Flood Washington with Leftist people. Prepare for revolution.

Train in the use of weapons of all sorts. Give the capitalists no rest! Harass them in their homes, businesses, restaurants and resorts! Carry signs and banners which call for the Dictatorship of the Proletariat and Council Power!

Above all else, make every effort to raise the level of awareness of the proletariat. The working class must be made aware of itself as a class!

We will know we are being successful when working are openly discussing Council Power and the Dictatorship of the Proletariat!

INDEX

Akie River, 115–118
Alternative facts, 123
American settlers, 104–106
Animals, allegedly extinct
    beaver-eating bear, 111–118
    dire wolf, 122
    Jefferson ground sloth, 120–121
    sabre-toothed cat, 123
    short-faced bear, 109–118
    woolly mammoth, 99–107
Anthropology, 3
Asteroid extinction theory, 6–7
Bering Strait migration, 33
Bush pilots, 107
Caribou, 41
Climate change, 95–97
Continental drift theory, 7
Cultural conflict, 22–23
Davis Creek, 112
Dene people
    basket-weaving, 24
    cultural traditions, 22–28
    Elders, 15–20
    hunting practices, 22, 111–118
    life in the mountains, 19–21, 29–36
    Sekani tribe, 13–15
Dire wolf, 122
Dinosaurs, extinction of, 4–7
Elephants, African, 104
Elders (Indigenous), 15–20, 109
Extinction theories, 5–7, 95–97
Forests, 16, 37
Glaciers, 95–97

Grizzly bear, 24, 41
Indigenous knowledge, 14–20, 29–36
Jefferson ground sloth, 120–121
Logging roads, 47–53
Mass extinctions (disputed), 5–7, 95–97
Milwaukee lion (sabre-toothed cat), 123
Muskeg, 100
Palaeontology, 2
Prairies, 34–35, 103–105
Railroads, 104–105
Rocky Mountain Trench, 13–15, 31–32
Sabre-toothed cat (Smilodon fatalis), 123
Sekani people, 13–15
Short-faced bear (Arctodus simus), 109–118
Stone Age parallels, 22
Swamps, 100
Tourism and wildlife, 106
Trapping, 111–118
Tsay Keh Dene (Mountain People), 13–15
Villages, life in, 19–21, 45–47
Wilderness wolf (Dire wolf), 122
Woolly mammoth, 99–107

Albatross (comparison), 142
Beast (Biblical), 136
Bible references
    Dragon in Africa, 134
    Revelations, 136

Bird from hell (Devil Bird / Pterosaur), 129–147
Caves, nesting sites, 139–141
Chameleon (camouflage reference), 150
Cliff-climbing behavior, 141–143
Dene Elders
  belief in Devil Bird, 129–131
  cultural descriptions, 133–135, 138–139
  warnings about sundown, 138
Devil Bird (see also *Pterosaurs*), 129–147
Devil Dog (pterosaur on all fours), 144
Dragon (mythological/biological link), 133–136
Eggs (pterosaur reproduction), 148–155
Flaplings (pterosaur young), 154–160
Hell, caves as gateway to, 140
Jersey Devil, 137
Leprechauns (flaplings as "Little People" in other cultures), 159–160
Little People (Dene name for flaplings), 158
Mosquitoes (food source), 156
Nesting grounds (caves in mountains), 139–141
Nesting sites (hilltops for eggs), 148–150
New Testament, Revelations, 136
Nocturnal habits (pterosaurs), 139, 147
Ostrich eggs (comparison), 152–154
Pterodactyls (short-tailed pterosaurs), 132–133
Pterosaurs
  as reptiles, 147–155
  fossilized trackways, 142–143
  lifting power, 137–138
  size and wingspan, 136–138
  walking on wings, 142–143
Radar detection (question of visibility), 146
Rhamphorynchus (long-tailed pterosaurs), 132–133
Satan (association with Devil Bird), 136, 140
Snapping sound (wing opening), 143
Swamps (nurseries for flaplings), 156–157
Thunder Bird, 137

Archaeopteryx (early bird), 172
ATP (biochemistry of bioluminescence), 185
Bats
  flight mechanics, 169–171
  competition with pterosaurs, 176
Bioluminescence (pterosaurs glowing), 183–188
Birds
  competition with pterosaurs, 165–168, 177
  flight mechanics, 172–175
  raptors (birds of prey), 166–167, 178
Caribou (antler comparison), 186–187
Christmas dinner (turkey vs. pterosaur), 168
Cloud of smoke (toxic gas), 192–194
Competition for skies
  birds vs. pterosaurs, 165–168
  raptors vs. flying reptiles, 177–178
Darwin (pre-evolution era science), 164
Death and mutilation of livestock, 190–197
Devil Bird (see also *Pterosaurs*), 202–210
Dinosaur (misuse of term), 163–165
Disappearance of girls of child-bearing age, 202–212
DNA testing (proving reptilian identity), 196
Dragons (fire-breathing legend), 194
Egg-layers
  birds vs. reptiles, 164
FBI (response to livestock deaths), 195
Flight mechanics

lift, wing curvature, 168–170
bird vs. reptile efficiency, 172–175
pterosaur shoulders and pivot joints, 171–173
Forest Service (logging road permits), 196
Fossilized trackways (bat-like walking of pterosaurs), 170
Gates of hell (cave entrances), 197
Gigantothermia (heat retention in reptiles), 213
Glow (pterosaur mating and hunting), 183–188
Highway of Tears (missing women), 203–206
Imitation sounds (dogs, babies, women screaming), 214–216
Leatherback turtle (comparison to gigantothermia), 213
Livestock deaths (see also *Death and mutilation of livestock*), 190–197
Luciferase / luciferin (biochemistry of glow), 185
Madison Scott (disappearance case), 210–211
Mass extinction theory (rejection of), 164–165
MMIW (Missing and Murdered Indigenous Women), 208–209
Modern aircraft (flaps analogy), 170–171
Noise
 flapping wings (whooshing), 207
 whistling (Thunder Bird), 210
Owls and night hawks (competition with pterosaurs at night), 178
Pangea (world landmass), 160–162
Poison gas (pterosaur weapon), 192–194
Predatory attacks (girls, children), 161, 202–204
Prince Rupert–Edmonton Highway (Highway of Tears), 203–205

Raptors (birds of prey), 166–168, 178
Reptiles vs. birds (cold-blooded vs. warm-blooded), 174–175
Road cameras (proof strategy), 196–197
Shoulder joints (pterosaur flight advantage), 171–172
Smoke (toxic cloud), 192–194
Sounds imitated by pterosaurs, 214–216
Ten Thousand Road (logging road to caves), 196–197
Thunder Bird (whistling cry), 210
Trail cameras (proposed proof), 196–197
UFO's / IFO's (glowing pterosaurs mistaken for), 183–185, 190
Whistling attack sound, 210
Women attacked (child-bearing age, menstruation scent), 161, 202–208

Apes
 transition to humans (bipedalism, opposable thumb), 245–247, 262–264
Bigfoot (see also Sasquatch; Giants), 244–255
Bipedalism (key to becoming human), 245–247, 262–264
Brain size
 not determinant of humanity, 259–261, 265
British Columbia (Indigenous Reserves, Sasquatch on beaches), 252–253
Burger, Lee (paleoanthropologist), 262–267
Cave burials (Rising Star system, Homo Naledi), 263–268
Childhood of the human race (Engels' concept), 264
Dené people (term "Stink People" for Sasquatch), 246–247

Dogs (fear, silence in presence of Giants), 249–250

Engels, Friedrich
  on ape-to-human transition, 246–247, 264–265
  Origin of the Family, Private Property and the State, 264

Europe (royalty, inbreeding as cautionary tale), 251

Flores "Hobbits" (Homo Floresiensis), 257–259

Food habits of Giants (seasonal migration, hunting-gathering), 248–249

Forth, Gregory (anthropologist, Flores humans), 258–259

Giant humans (see also Sasquatch; Bigfoot), 244–255

Gigantopithecus (as human species), 244–247

Grave robbing (concerns with Homo Naledi remains), 268–269

Hobbits (see Flores "Hobbits")

Homo Naledi
  discovery in Rising Star caves, 262–267
  burial practices, 263–265
  human status debated, 262–267

Homo sapiens (not the only human species), 244–246, 257

Hunting of Sasquatch (condemned as murder), 247–248

Indigenous Elders (role in Sasquatch contact), 252–253

Indigenous peoples (trust with Giants), 252–253

Inbreeding (dangers for Giants), 250–252

Island of Flores (Homo Floresiensis), 257–259

Marxist understanding of evolution, 246–247, 264–265

Mirrors, ornaments, gifts for Giants, 253–254

Morgan (anthropologist, quoted by Engels), 266

Naledi (see Homo Naledi)

Opposable thumb (definition of human), 245–247, 262–264

Origin of the Family, Private Property and the State (Engels), 264

Paleoanthropology, 262–263

Rising Star cave system (South Africa), 262–268

Royal family (England, inbreeding example), 251

Sasquatch (see also Bigfoot; Giants), 244–255

Seasonal migration of Giants (food supply), 248–249

Silence of dogs (in Sasquatch presence), 249–250

Stink People (Dené term for Sasquatch), 246–247

Tooting (flatulence as insult response), 246–247

United Nations (laws to protect Giants), 255

"Universal, non-verbal display of contempt" (flatulence), 246–247

- Ambergris, value of – 142
- Apex predators, ecological role of – 133
- Basilosaurus (freshwater whale) – 128
- Cave-dwelling behaviour – 139
- Drones, use of in searches – 150
- Ecosystem comparisons (wolves, hippos, orcas) – 135
- Eyewitness reports (Ogopogo, Loch Ness) – 146
- Hippopotamus, comparison with – 136

- Ichthyosaurs, reclassified as mammals — 130
- Inbreeding, danger of — 138
- Lake monsters, feeding habits — 144
- Loch Ness Monster, historical and modern accounts — 147
- Mosasaurs, live birth and survival — 131
- Ogopogo, sightings and searches — 145
- Plesiosaurs, mammalian traits — 129
- Predatory risks to humans — 141
- Proletarian science, importance of — 154
- Reward for proof of Ogopogo — 148
- Trail cameras, placement of — 151
- Underwater lights, possible bioluminescence — 152
- Whale evolution and classification — 132

- Academic conformity / memorization vs. critique — 156
- Adelson/Alvarez (meteor mass-extinction theory) — 162
- Astronomy — historical challenges (Copernicus, Kepler, Galileo) — 158
- Career suicide (scientists who challenge orthodoxy) — 157
- Climate change / global warming (discussion) — 170
- Copernican revolution (example of scientific challenge) — 158
- Dictatorship of the Proletariat (political tie-ins) — 176
- Engaging workers in science (practical cooperation) — 171
- Galilean method / scientific method (proper) — 159
- Global warming: natural climate variability — 171

- Hobbits / Homo floresiensis (scientific dissent and support) — 160
- Kepler / Newton / standing on shoulders of giants — 158–159
- Loch Ness Expedition (scientists taking risks) — 161
- Mass extinction of dinosaurs (critique) — 162–163
- Middle Ages analogy (retrograde science) — 156–157
- Moon landing skepticism (Van Allen belts argument) — 168–169
- Monopolies, capitalism and science (impact on research) — 174
- Pterosaurs & sacred theories (not to be challenged) — 164
- Professional advice to scientists & officials — 175
- Proletarian science / theoretical struggle (Lenin, Engels, Marx) — 173–174
- Preparing working class for council power (call to action) — 176
- Retrograde trend in science (root causes) — 156–157
- Scientific heresy: risks and examples — 157–162
- Universities (bourgeois distortions of Marxism) — 172

Animals & Legends
- Dragons / Thunder Birds — 115
- Gigantopithecus (Giants / Sasquatch / Bigfoot) — 118
- Ogopogo (Lake Monster) — 121
- Pterosaurs (Flying reptiles) — 115
- Predation on humans & livestock — 116

Approach & Proof
- DNA and toxin testing of livestock carcasses – 117
- Indigenous Elders, cooperation with – 119
- Gifts for Giants (food, mirrors, cosmetics, burlap bags) – 120
- Trail cameras (for Ogopogo & others) – 121

Locations
- Buffalo Head Mountains (pterosaur nesting grounds) – 117
- Okanagan Lake (Ogopogo) – 121
- Pacific Coast Reserves (Giants' beaches) – 119
- Ten Thousand Road (logging access to caves) – 117

Political Context
- Bourgeois schools (critique by Lenin) – 114
- Councils (Soviets) and revolutionary organizing – 122
- Dictatorship of the Proletariat – 122
- Role of workers & farmers in discoveries – 114
- Training workers for authority after revolution – 123

Historical & Revolutionary References
- American Revolution (tar and feathering, breaking Tory spiritual power) – 132–134
- Chinese Revolution (landlords, dunce caps, humiliation rituals) – 131–133
- Lenin (1918 speech on education, *State and Revolution*) – 139–141
- Marx: *Civil War in France* (Paris Commune, smashing state machine) – 125–127
- Marx & Engels, *Communist Manifesto* (1847, proletariat as ruling class) – 124
- Paris Commune (1871, parson power, smashing state) – 125–127

Spiritual Forces of Repression
- Clergy, "Parson Power" – 126
- Invisible chains (spiritual power as repression) – 126–127, 131
- Landlords (China, humiliation breaking power) – 131–132
- Mobsters / Organized crime (extortion, humiliation as liberation) – 134–137
- Professor Power (scientists, professors, education system) – 137–146
- Spiritual repression in education (school textbooks, bourgeois ideology) – 138–141

Science & "Sacred" Theories
- Dinosaurs (misclassification, extinction debate) – 142–145
- Gigantopithecus / Sasquatch (denial by scientists) – 143
- Ichthyosaurs & Basilosaurus (Loch Ness Monster, Ogopogo, "walking whales") – 145–147
- Mass extinction theories (dinosaurs, megafauna, climate change myths) – 143–144
- UFOs / bioluminescent pterosaurs – 144
- Whales, land-walking – 145

Political & Philosophical Themes
- Capitalist class (imperialism, reactionary nature) – 147–150
- Class traitors (bootlickers, loyalty to capitalists) – 153–154
- Dictatorship of the Proletariat (necessity, suppression of enemies) – 135, 140–141, 148–149

- Education as bourgeois tool (obedient lackeys, false knowledge) – 138–141
- Fall & decline of civilizations (imperialism driving collapse) – 146–148
- Monopoly capitalists / billionaires (enemies of proletariat) – 147–149, 152–153
- Soviet (Council) Power – 141, 150
- Spiritual vs physical repression (distinction, both must be destroyed) – 154–155

Warnings & Revolutionary Lessons
- Failure to break spiritual power (China after Mao, capitalist return) – 155
- Need to smash both physical and spiritual state apparatus – 154–155
- Role of advanced workers (training for Dictatorship of Proletariat) – 141–142
- Working-class tribunals, public humiliation – 134–136

Animals & "Sacred Theories"
- Dragons / Thunder Birds (pterosaurs, predation, Ten Thousand Road caves) – 201–205
- Giants / Gigantopithecus / Sasquatch – 200–202
- Ichthyosaurs & Basilosaurus (Loch Ness, lake monsters, walking whales) – 198–203
- Lake Monsters (Ogopogo, Champie, Bessie, Great Lakes) – 196–199
- Pterosaurs (poison gas, reptilian DNA) – 204–205
- Trail cameras, drones, & proof of existence – 197–198

Working Class & Class Awareness
- Awareness of class interests – 187–190
- Bourgeoisie (billionaires, monopoly capitalists) – 187–188, 210–212
- Common people / proletarians / "little guy" – 186–188
- Petty bourgeois (small business, professionals as allies) – 186–187
- Raising self-esteem through scientific discovery – 195–197
- Workers' training through field exercises – 196–198, 204

Marxist Theory & Revolutionary Texts
- Engels: three great struggles of Communism – 192
- Lenin: *State and Revolution* – 188–190, 206–207
- Lenin: *What Is To Be Done?* – 191–193
- Lenin: *Left-Wing Communism: An Infantile Disorder* – 208
- Marx & Engels: *Communist Manifesto* – 190
- Revolutionary theory and class consciousness – 189–192

Political Divisions & Tendencies
- Bolsheviks (majority, Lenin's leadership) – 192
- Mensheviks (minority, revisionists, opportunists) – 192–193
- Opportunism / "Economism" – 193–194
- Revisionists / Social chauvinists – 190–192
- Socialists (non-Marxist allies of Communists) – 194

Councils / Soviets
- Councils as revolutionary organs – 200–202, 210–212
- First appearance in Russia, 1905 – 200

- Seattle Autonomous Zone (modern parallel) – 201
- Soviets as national authority in insurrection – 211–212

Revolution & Insurrection
- Insurrection (seizing key locations, transport, communications) – 206–208
- Necessity of insurrection vs. strikes – 206–207
- Strikes, wildcat strikes, slowdowns – 206
- Training for insurrection through scientific proof – 196–197, 204
- January 6, Washington D.C. comparison – 206
- Zones (Seattle, Occupy Movement) – 201

Comparisons to Russian Revolution
- April return of Lenin – 208–209
- Revolutionary July Days (aborted insurrection) – 209
- October Revolution (November 7 new style) – 209–210
- Consciousness of Russian vs. American proletariat – 209–210

Communist Party & Leadership
- Creation of true Communist Party – 211–212
- Councils vs. Communist Party (roles, cooperation) – 211–212
- Leadership of middle-class intellectuals – 210
- Party as national authority during insurrection – 212
- Slogan: "Prepare for Council Power and Dictatorship of the Proletariat" – 212

Political Context & Crisis
- Billionaires (crisis of rule, fear of revolution) – 220–223
- Great Depression comparison ("on steroids") – 227–228
- Middle class impoverishment, collapse of small business – 226–228
- Trump Senate trial (political turning point) – 219–220
- White House Resistance (capitalist crisis) – 220

Scientists & Intellectuals
- Association of Scientists (proposed international organization) – 222–224
- Career risk & protection in numbers – 222
- Internet as revolutionary tool (dissemination of Marx & Lenin) – 224–225
- Middle class intellectuals in Communist Party formation – 229–231
- Role of scientists in exposing capitalist lies (textbooks, false science) – 221–222
- Scientists distancing from billionaires (safety during revolution) – 231–232

Revolutionary Theory & Leadership
- Councils (Soviets) – 225, 229
- Communist Party, creation of – 229–231
- Dictatorship of the Proletariat – 225, 230–232
- Marx and Lenin as political scientists – 223
- Proletariat as revolutionary class, "Act of God" – 227–228
- Raising awareness of working class – 219, 231

Slogans & Calls to Action
- "Workers of the World, Unite!" – 232
- "Scientists of the World, Unite!" – 232
- "Scientific Socialism!" – 232

- "Dictatorship of the Proletariat!" – 232

Science & Extinction Theories
- Creationism vs. evolution – 235–236
- Decline in science under monopoly capitalism – 240–243
- Evolution (Darwin's contribution to biology) – 235–236
- Mass extinction theories, rejection of – 241–242
- Newton's three laws of motion – 234–235
- Pterosaurs, predation on humans – 243–244
- Suppression of scientific dissent (career suicide, blackballing) – 241–242

Religion & Ideology
- Bible as source of all answers (pre-Newton worldview) – 234
- Creationists & fundamentalists (opposition to evolution) – 236
- Reactionaries vs. fundamentalists (political vs. religious resistance) – 237

Civilization & Decline
- Civilizations: rise, peak, and decline – 237–238
- Decline of present civilization (caused by reactionaries) – 237–239
- Industrial Revolution (creation of bourgeoisie and proletariat) – 238–239
- Role of reactionaries in destruction of civilization – 238
- Sacrifices of ancestors in building present civilization – 237–238

Political Economy
- Bourgeoisie: revolutionary role in early capitalism – 238–239
- Capitalism: dual nature (progressive & reactionary aspects) – 248–249
- Imperialism (monopoly stage of capitalism) – 239–240, 248
- Lenin: *Imperialism, the Highest Stage of Capitalism* – 240
- Proletariat as revolutionary class – 239–240, 249
- Student debt, tuition barriers, and class effects on science – 241–242

Democracy & Rights
- Declaration of Independence (right to abolish government) – 246–247
- Democratic rights under attack – 244–247
- Founding Fathers and democracy – 246–247
- Plutocracy of the "1%" (Citibank memo, Occupy movement) – 244–246
- U.S. Constitution, silence on capitalism – 246

Culture & Media
- Jefferson and Washington, glorification of slave owners – 242–243
- Michael Moore, *Capitalism: A Love Story* – 244–245
- Occupy Movement (99% vs. 1%) – 244–246

Revolutionary Theory
- Class struggle as history's driving force – 248
- Communist Manifesto (quotes, bourgeoisie and proletariat) – 238–239, 248
- Engels' 1883 introduction – 248
- Dictatorship of the Proletariat – 249–250
- Scientific Socialism vs. Utopian Socialism – 249
- Working class as revolutionary force – 239–240, 249–250

www.ingramcontent.com/pod-product-compliance
Lightning Source LLC
Chambersburg PA
CBHW020459030426
42337CB00011B/158